Lukas Chávez

## Epigenetic Analysis of Human Embryonic Stem Cells

Lukas Chávez

# Epigenetic Analysis of Human Embryonic Stem Cells

## Basics, Computational Methods, Applications

Südwestdeutscher Verlag für Hochschulschriften

**Impressum/Imprint (nur für Deutschland/only for Germany)**
Bibliografische Information der Deutschen Nationalbibliothek: Die Deutsche Nationalbibliothek verzeichnet diese Publikation in der Deutschen Nationalbibliografie; detaillierte bibliografische Daten sind im Internet über http://dnb.d-nb.de abrufbar.
Alle in diesem Buch genannten Marken und Produktnamen unterliegen warenzeichen-, marken- oder patentrechtlichem Schutz bzw. sind Warenzeichen oder eingetragene Warenzeichen der jeweiligen Inhaber. Die Wiedergabe von Marken, Produktnamen, Gebrauchsnamen, Handelsnamen, Warenbezeichnungen u.s.w. in diesem Werk berechtigt auch ohne besondere Kennzeichnung nicht zu der Annahme, dass solche Namen im Sinne der Warenzeichen- und Markenschutzgesetzgebung als frei zu betrachten wären und daher von jedermann benutzt werden dürften.

Coverbild: www.ingimage.com

Verlag: Südwestdeutscher Verlag für Hochschulschriften GmbH & Co. KG
Dudweiler Landstr. 99, 66123 Saarbrücken, Deutschland
Telefon +49 681 37 20 271-1, Telefax +49 681 37 20 271-0
Email: info@svh-verlag.de

Approved by: Berlin, FU, Diss., 2010

Herstellung in Deutschland:
Schaltungsdienst Lange o.H.G., Berlin
Books on Demand GmbH, Norderstedt
Reha GmbH, Saarbrücken
Amazon Distribution GmbH, Leipzig
**ISBN: 978-3-8381-2762-0**

**Imprint (only for USA, GB)**
Bibliographic information published by the Deutsche Nationalbibliothek: The Deutsche Nationalbibliothek lists this publication in the Deutsche Nationalbibliografie; detailed bibliographic data are available in the Internet at http://dnb.d-nb.de.
Any brand names and product names mentioned in this book are subject to trademark, brand or patent protection and are trademarks or registered trademarks of their respective holders. The use of brand names, product names, common names, trade names, product descriptions etc. even without a particular marking in this works is in no way to be construed to mean that such names may be regarded as unrestricted in respect of trademark and brand protection legislation and could thus be used by anyone.

Cover image: www.ingimage.com

Publisher: Südwestdeutscher Verlag für Hochschulschriften GmbH & Co. KG
Dudweiler Landstr. 99, 66123 Saarbrücken, Germany
Phone +49 681 37 20 271-1, Fax +49 681 37 20 271-0
Email: info@svh-verlag.de

Printed in the U.S.A.
Printed in the U.K. by (see last page)
**ISBN: 978-3-8381-2762-0**

Copyright © 2011 by the author and Südwestdeutscher Verlag für Hochschulschriften GmbH & Co. KG and licensors
All rights reserved. Saarbrücken 2011

For my family.

# Contents

**1 Introduction**   **3**
    1.1 Human embryonic stem cells (hESCs) . . . . . . . . . . . . . . . . 3
    1.2 Transcriptional regulation of pluripotency . . . . . . . . . . . . . . 5
    1.3 DNA methylation . . . . . . . . . . . . . . . . . . . . . . . . . . . 8
    1.4 Introduction to methods and limitations for detecting DNA methylation . 9
    1.5 Aims of the book . . . . . . . . . . . . . . . . . . . . . . . . . . . 11

**2 Experimental techniques**   **16**
    2.1 DNA arrays . . . . . . . . . . . . . . . . . . . . . . . . . . . . . . 17
    2.2 Statistical testing of differential expression . . . . . . . . . . . . . . 20
    2.3 Second generation sequencing . . . . . . . . . . . . . . . . . . . . . 25
    2.4 Chromatin immunoprecipitation (ChIP) . . . . . . . . . . . . . . . 27
    2.5 Methylated DNA immunoprecipitation (MeDIP) . . . . . . . . . . . 28
    2.6 Alternative methods for detecting DNA methylation . . . . . . . . . 30

**3 Modelling of MeDIP-seq data**   **33**
    3.1 Genome vector . . . . . . . . . . . . . . . . . . . . . . . . . . . . . 33
    3.2 Quality control metrics . . . . . . . . . . . . . . . . . . . . . . . . 38
        3.2.1 Saturation analysis . . . . . . . . . . . . . . . . . . . . . . . 38
        3.2.2 Coverage analysis . . . . . . . . . . . . . . . . . . . . . . . 45
        3.2.3 CpG enrichment . . . . . . . . . . . . . . . . . . . . . . . . 48
    3.3 Normalization . . . . . . . . . . . . . . . . . . . . . . . . . . . . . 50
        3.3.1 Reads per million (rpm) . . . . . . . . . . . . . . . . . . . . 51
        3.3.2 Coupling factors . . . . . . . . . . . . . . . . . . . . . . . . 51
        3.3.3 Calibration curve . . . . . . . . . . . . . . . . . . . . . . . . 55
        3.3.4 Relative methylation score (rms) . . . . . . . . . . . . . . . 59

*Contents*

|  |  | 3.3.5 | Absolute methylation score (ams) | 64 |
|---|---|---|---|---|
|  |  | 3.3.6 | Evaluation of MeDIP-seq data normalization | 65 |
|  | 3.4 | | Identification of differentially methylated regions (DMRs) | 70 |
|  | 3.5 | | MEDIPS software package | 74 |

**4  Epigenetic dependencies during endodermal differentiation of human ES cells  77**

|  | 4.1 | Identifying a core regulatory network of OCT4 controlling pluripotency | 78 |
|---|---|---|---|
|  | 4.2 | Experimental procedures and analysis methods | 81 |
|  | 4.3 | MeDIP-seq quality control metrics | 84 |
|  | 4.4 | Comparing MeDIP-seq and WGSBS derived methylation profiles | 87 |
|  | 4.5 | Promoter methylation | 88 |
|  | 4.6 | Methylation patterns of transcription factor binding sites | 90 |
|  | 4.7 | Identification of differentially methylated regions (DMRs) | 90 |
|  | 4.8 | Genome wide distribution of DMRs | 93 |
|  | 4.9 | Differential methylation at TFBSs | 94 |
|  | 4.10 | Enrichment analysis associates de-methylation events to functional histone modifications | 96 |
|  | 4.11 | Differential methylation and gene expression alterations | 98 |
|  | 4.12 | Epigenetic effects on the OCT4 regulatory network | 101 |

**5  Conclusion**     **104**

**Bibliography**     **109**

**Notation and Abbreviations**     **123**

**Publications**     **124**

**Acknowledgments**     **126**

**Appendix 1**     **128**

# 1 Introduction

## 1.1 Human embryonic stem cells (hESCs)

In 1989, Thomson and colleagues presented for the first time the derivation of hESCs from the inner cell mass of human blastocysts [98]. For this, cleavage stage human embryos, produced by *in vitro* fertilization for clinical purposes, were cultured to blastocyst stage and the inner cell masses were isolated (see Figure 1.1). Thomson and colleagues defined three essential characteristics of primate embryonic stem cells [98]. These are (i) derivation from the preimplantation or periimplantation embryo, (ii) prolonged undifferentiated proliferation (*stemness*), and (iii) stable developmental potential to form derivatives of all three embryonic germ layers even after prolonged culture (*pluripotency*). Human ESCs offer insights into developmental events which cannot be studied directly in the intact human embryo but that have important consequences in clinical areas. For example, understanding the mechanisms that control differentiation enables directed differentiation of hES cells to specific cell types [98].

However, derivation of ESCs from human blastocysts for clinical applications has to be considered critical because of ethical and technical reasons. The ethical concerns are obvious, because isolation of the inner cell mass by destruction of the embryo inversely leads to the abortion of embryogenesis. From the technical point of view, there are two main difficulties for putative future cell-replacement therapies. First, for putative future clinical applications, hESCs are not allowed to be contaminated by reagents used in cell culture. Second, donor-host rejections are likely to arise caused by differing genomic backgrounds of patient and cell line. Fortunately, alternative approaches for derivation of hES and hES-like cells have been recently developed.

# 1 Introduction

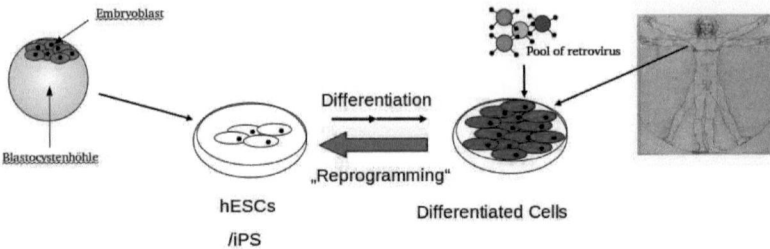

**Figure 1.1: Embryonic stem cells and reprogramming.** Embryonic stem cells are derived from the inner cell mass of mammalian blastocysts. They have the developmental potential to differentiate into derivatives of all three embryonic germ layers even after prolonged culture (*pluripotency*). Several studies on reprogramming somatic cells to induced pluripotent stem cells (iPSCs) have demonstrated that the transduction of only a few transcription factors (TFs) is sufficient for resetting differentiated cells into a molecular state similar to embryonic stem cells.

On the one hand, Kilmanskaya and colleagues [50] have shown that hESCs can be derived from single blastomers without interfering with the embryonic developmental potential. On the other hand, several studies on reprogramming somatic cells to induced pluripotent stem cells (iPSCs) have demonstrated that transduction of only a few transcription factors (TFs) is sufficient for resetting differentiated cells into a molecular state similar to embryonic stem cells (see Figure 1.1). In 2006, Takahashi and colleagues [96] have demonstrated for the first time that iPSCs can be generated from adult fibroblasts in mouse. Subsequently, in 2007, two independent laboratories have presented human iPSCs derived from differentiated fibroblasts [95, 107]. Both groups have shown that their iPS cells fulfill the *pluripotency* characteristic of hESCs. For this, they show iPSCs differentiation into derivatives of all three primary germ layers. Furthermore, iPSCs resemble hESCs in their morphology, gene expression, chromatin state and DNA methylation characteristics [104].

Taken together, these studies demonstrated that hESCs or iPSCs can be generated without being burdened by the ethical concerns of destructive hESC generation. The possibility to generate patient specific iPSCs will putatively reduce donor-host rejections in future cell-replacement therapies. Nevertheless, the utilization of transfected somatic cells for clinical applications has still to be considered critical, because of possible undesirable side effects. However, for research, hESCs and iPSCs already serve as a valuable resource for analyzing the molecular mechanisms responsible for targeted differentiation into functional cell types.

## 1.2 Transcriptional regulation of pluripotency

The POU domain transcription factor OCT4 is highly expressed in the inner cell mass of mammalian blastocysts. It is supposed that pluripotency and self-renewal is controlled by a transcription regulatory network governed by OCT4 and that down-regulation of OCT4 is crucial for initiating early embryonic differentiation [6].

Several studies on reprogramming human somatic cells to iPSCs even emphasize the role of OCT4 as key regulator of pluripotency in the early mammalian embryo. Here, the transduction of only a few transcription factors is sufficient for resetting differentiated cells into a molecular state similar to embryonic stem cells. While Takahashi et al. [95] and Wernig et al. [104] obtained iPS cells by transduction of the TFs OCT4, SOX2, KLF4, and C-MYC, Yu et al. [107] achieved similar results with a transcription factor set composed of OCT4, SOX2, NANOG, and LIN28. Interestingly, only the TFs OCT4 and SOX2 are common in both approaches. Moreover, Huangfu et al. [39] demonstrated that iPSCs can be derived at higher efficiencies by the transduction of these two factors in combination with histone deacetylase inhibitor -valproic acid.

The TF OCT4 is known as a key regulator for maintaining pluripotency in mammalian embryos [73, 81, 1] and is exclusively expressed in mammalian embryonic stem cells. For example, it was shown that OCT4 is expressed neither in functional somatic cells nor in adult stem cells, by a microarray based gene expression profile comparison between hESCs, adult liver progenitor cells, and hepatocytes [48]. The HMG-box containing TF SOX2 interacts with OCT4 and the SOX2/OCT4 heterodimer complex is able to promote selective gene activation or repression during mammalian embryogenesis [82, 12, 74].

Functional data on OCT4 regulatory action is available from heterogeneous sources: to reveal DNA-Protein binding events of OCT4, SOX2 and of the pluripotency associated TF NANOG, chromatin immuno-precipitation followed by microarray experiments

# 1 Introduction

(ChIP-on-chip) has been performed using hESCs [13]. Complementary to OCT4 binding events in hESCs, we have identified OCT4 binding events in human embryonic carcinoma cells by ChIP-on-Chip experiments [49]. Moreover, OCT4 specific sequence motifs have been identified. Two examples are the octamer motif $ATTTGCAT$ that interacts with POU domain factors like the homeodomain containing TF OCT4, as well as a motif recognized by the SOX2/OCT4 heterodimer complex [93, 18, 87]. Mapping of these known transcription factor binding motifs to the promoter sequences of putative OCT4 target genes provides additional evidence for direct binding events.

Although ChIP-on-chip experiments and sequence-based methods have the ability to detect putative protein-DNA binding sites, these techniques do not allow inference of directional transcriptional dependencies between DNA binding and the effect on gene expression regulation. In order to test the regulatory influence of OCT4 to the transcription rate of its target genes, Babaie et al. [6] performed RNA interference-mediated suppression of OCT4 function in the H1 hESC line and analyzed the resulting global gene expression changes by microarray experiments. Transcriptional changes induced by OCT4 knockdown are expected to include genes linked with pluripotency, and genes activated upon differentiation along the trophoblast lineage [6].

ChIP-on-chip experiments, promoter sequence analysis and RNA interference provide complementary pieces of information on transcriptional dependencies. In order to identify a core OCT4 regulatory network in hESCs, I have performed [17] an integrated analysis of such high-throughput data along with promoter sequence analysis (see section 4.1). The resulting core OCT4 regulatory network controlling pluripotency in hESCs is composed of the set of genes detected by each of the individual experimental apprsoaches (see Figure 1.2).

Taken together, pluripotency is governed by the TF OCT4 and critical regulatory acting protein-DNA interactions have been investigated. Beside these transcription factor mediated regulations, it is supposed that there exist epigenetic mechanisms involved in regulation of stem cell differentiation and development, including histone modifications and DNA methylation. However, such epigentic mechanisms of hESC differentiation have

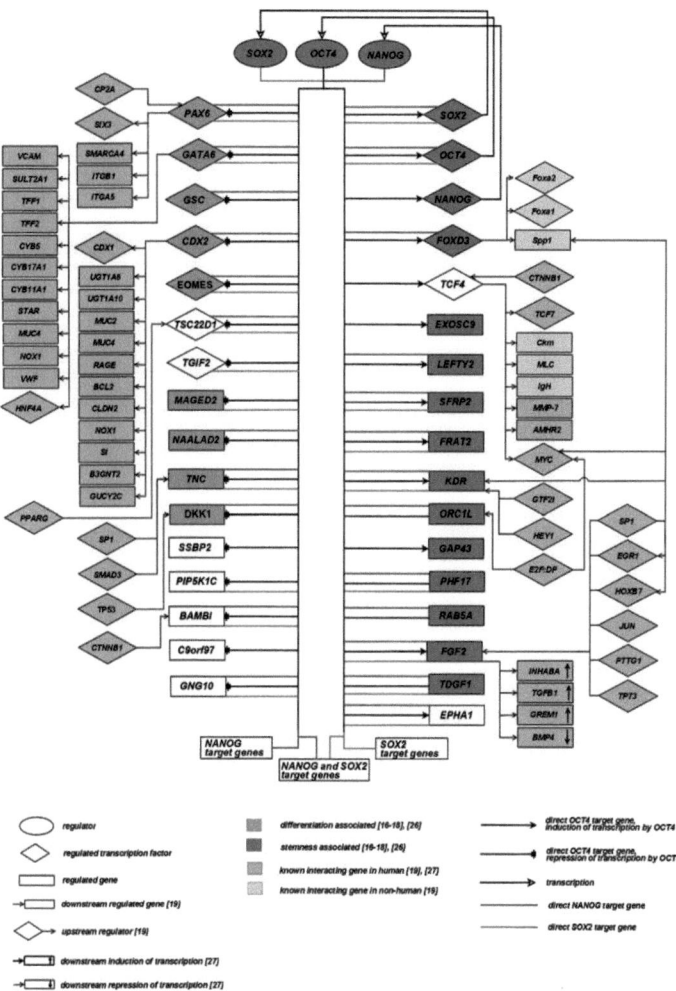

**Figure 1.2: OCT4 core regulatory network.** Core OCT4 transcriptional regulatory network identified by an integrative analysis of ChIP-on-Chip, RNAi knockdown and sequence-based motif mapping [17]. Green boxes represent genes associated with differentiation, and red boxes indicate genes being specific for hESCs [6, 2, 75, 5]. The network also incorporates information on direct target genes from SOX2 (red lines) and NANOG (blue lines) ChIP-on-chip experiments and an additional level of gene regulation [68, 34].

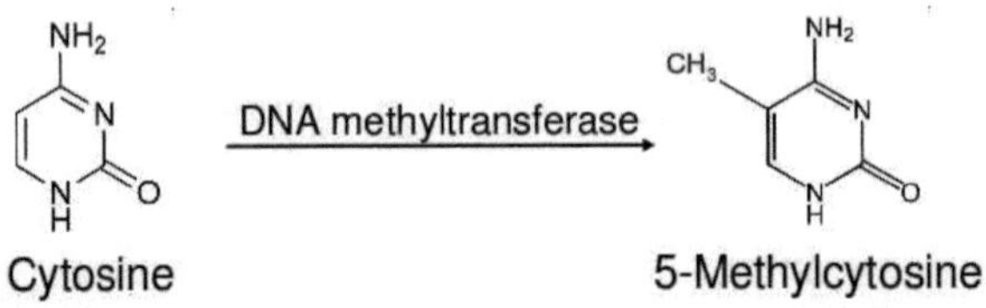

**Figure 1.3: Cytosine methylation.** DNA methyltransferases mediate the reversible attachment of a methyl group (CH$_3$) at the 5' position of cytosines.

not yet been investigated on a full genome level. This work reveals for the first time global epigentic modifications emerging upon induced loss of pluripotency and during differentiation of hESCs along the endodermal lineage. Moreover, important interactions between DNA methylation, transcription factor binding, and gene expression alterations are revealed. Finally, epigenetic dependencies during endodermal differentiation are presented in detail for the core OCT4 network that controls pluripotency.

## 1.3 DNA methylation

In mammals, DNA methylation describes the reversible attachment of a methyl group (CH$_3$) at the 5' position of the pyrimidine derivate cytosine. DNA methylation does not change the DNA sequence itself, and is therefore considered as an epigenetic modification. The process of DNA methylation is mediated by methyltransferase enzymes (see Figure 1.3).

Methyltransferase enzymes are commonly classified according to their enzymatic activities. On the one hand, maintenance of methylation activity is necessary to preserve DNA methylation after every cellular DNA replication cycle. The proposed maintenance methyltransferase responsible for copying DNA methylation patterns to the daughter strands during DNA replication is DNMT1 [43]. On the other hand, *de novo* methylation sets up DNA methylation patterns in early embryonic development. It is thought that DNMT3a and DNMT3b are the methyltransferases responsible for this *de novo* methylation. Moreover, DNMT3L is a protein which is homolog to other DNMT3s but has no direct catalytic activity. Instead, DNMT3L assists the *de novo* methyltransferases by increasing their ability to bind to DNA and stimulating their activity [43].

# 1 Introduction

DNA methylation is involved in transcriptional regulation during embryonic differentiation [72] and reprogramming of somatic cells into induced pluripotent stem cells [26, 16]. Aberrant methylation can be associated with severe effects, for example the induction of cancer [47, 41]. Furthermore, distinct genome wide methylation patterns distinguish different cell-types [30, 85]. In mammals, DNA methylation primarily occurs at CpG sites but a recent study has shown that non-CpG methylation accounts for approximately 25% of all methylated cytosines in human embryonic stem cells [64]. It is supposed that methylation affects gene expression by either interfering with binding of transcription factors or modifying chromatin structure to a repressive state [43].

Meissner et al. [72] have shown that methylated CpGs are dynamic epigenetic marks that undergo extensive changes during cellular differentiation, particularly in regulatory regions outside of core promoters. Their data support the notion that both CpG-rich and -poor regulatory elements undergo distinct modes of epigenetic regulation. Regulatory elements in CpG-poor sequence contexts seem to undergo extensive and dynamic methylation and de-methylation. Hence, their study confirms the concept of distinct regulatory mechanisms of DNA methylation, where methylation of isolated CpGs may contribute to chromatin condensation or directly interfere with transcription factor binding. Moreover, hypermethylation of high-CpG-density promoters (HCPs) leads to irreversible gene silencing [72]. According to these findings, Rakyan et al. [85] observed a negative correlation between DNA methylation and gene expression at high-, but also at medium-, and contrary to previous notions, at even some low-CpG density promoters. On the other hand, Rakyan et al. observed that gene-body methylation positively correlates with gene expression [85].

## 1.4 Introduction to methods and limitations for detecting DNA methylation

In principle, there are two major high-thoughput methods for detecting DNA methylation. On the one hand, bisulphite based methods can be applied in order to produce DNA methylation information at base resolution. On the other hand, immunoprecipita-

## 1 Introduction

tion based methods are more cost-effective but received methylation levels are of lower resolution. For a detailed description of the individual techniques that will be mentioned in this section, see chapter 2.

Bisulphite sequencing or whole genome shotgun bisulphite sequencing (WGSBS) detect cytosine methylation on a base-pair level. Although whole genome single-base resolution maps have been generated [63, 64] such techniques cannot yet be cost-effective applied to screen large sets of sequences or samples. As an example, the first full genome methylome of hESCs on base resolution were recently reported at a cost of about 1.2 billion sequence reads [64]. Reduced representation bisulphite sequencing (RRBS) [72] address this issue by reducing the amount of DNA to be sequenced on the cost of a reduced view of genome wide methylation profiles.

Methylated DNA immunoprecipitation (MeDIP) uses an antibody specific for methylated cytosines in order to immunocapture methylated genomic fragments [103]. Immunoprecipitated methylated DNA fragments can be detected either by tiling arrays (MeDIP-Chip) or by next-generation sequencing (MeDIP-seq). Methylation profiles obtained by the MeDIP approach are not base-pair specific but reflect methylation levels on a resolution restricted by the size of the sonicated DNA-fragments after amplification and size selection. However, in contrast to WGSBS or RRBS, the MeDIP approach can be applied in order to obtain cost-effective and full-genome methylation levels without the limitations caused by methylation-sensitive restriction enzymes.

Nevertheless, it has been shown that MeDIP derived data needs to be corrected for local CpG densities in order to estimate valid methylation levels [27, 80]. This effect is caused by varying efficiency of antibody binding and immunoprecipitation dependent on the local density of methylated CpG sites. Especially the analysis of CpG-poor regions has been assumed to be difficult [103, 27]. Moreover, the use of tiling arrays or second generation sequencing for the detection of the immunoprecipitated methylated DNA fragments introduce platform dependent effects that have to be considered additionally. While there is applicable software available for analyzing MeDIP-Chip data [27, 80], normalization of MeDIP-seq data is in principle solved [27] but remains disproportionally time-consuming.

*1 Introduction*

In fact, processing of MeDIP-seq data from only one full chromosome (i.e. the human chromosome 1) takes approximately three days on a modern-day server when the BATMAN software [27] is applied.

Therefore, the major bottleneck of MeDIP-seq based methylation analysis is the time efficient processing of sequencing data with respect to its inherent complexity. There are several important but not yet well researched aspects in the context of MeDIP-seq data analysis including quality control metrics and data normalization. For example, the number of sequencing reads necessary for obtaining a sufficiently covered methylome relative to the size of the genome of interest has to be estimated. Second, the coverage of genome wide CpGs by the available data has to be examined. Third, enrichment of CpG rich short reads relative to the genomic background has to be analysed in order to provide a quality measure for antibody binding and immunoprecipitation. Moreover, there is the dire need for a time-efficient method that corrects for the DNA sequence specific bias introduced by the MeDIP experiment. Finally, there is no MeDIP-seq specific method that identifies events of differential methylation when different conditions are compared. This book presents novel concepts and time-efficient implementations targeting all of the mentioned existing limitations of MeDIP-seq based DNA methylation analysis, including quality control metrics, normalization, and identification of differential methylation on a full genome level.

## 1.5 Aims of the book

Generation of genome-wide data derived from methylated DNA immunoprecipitation followed by sequencing (MeDIP-seq) has become a major tool for epigenetic studies in health and disease. Computational analysis of such data, however, still falls short on accuracy, sensitivity and speed. In fact, the current bottleneck resulting from advancing technology development in DNA methylation is the computational analysis of the large-scale sequencing data [54] and, therefore, efficient experiment specific data analysis methods must be developed.

It has been previously shown that MeDIP-derived data need to be corrected for local

# 1 Introduction

CpG densities in order to compute unbiased methylation levels [27, 80]. This effect is caused by a varying efficiency of antibody binding and immunoprecipitation dependent on the local density of methylated CpG sites. Although there are computational methods available for analyzing MeDIP-Chip data [27, 80], in particular the normalization of MeDIP-seq data remains disproportional time-consuming. Moreover, important features for the design of MeDIP-seq experiments have not yet been addressed sufficiently, for example estimation of the number of reads necessary for obtaining a sufficiently covered methylome, the analysis of genome wide covered CpGs, the enrichment of CpG rich short reads relative to the genomic background, as well as statistical identification of differentially methylated regions between different conditions.

Here, I present the first method able to cope with the inherent complexity of MeDIP-seq data and outperforms computation time of existing methods by orders of magnitude with similar performance. As a proof of performance, I processed the available MeDIP-seq sperm data from Down et al. [27], compared the results to benchmark data from the HEP project [30], and show comparable concordance to the results of Down et asl [27]. In order to demonstrate the computational approach, I have analysed alterations in DNA methylation during differentiation of hESCs to definitive endoderm. I show improved correlation of normalized MeDIP-seq data in comparison to available whole-genome bisulphite sequencing data [64] and investigated the effect of differential methylation on gene expression. Furthermore, I analyzed the interplay between DNA- methylation, histone modifications, transcription factor binding, and show that in contrast to *de-novo* methylation, de-methylation is mainly associated with regions of low CpG densities.

In chapter 2, I describe different experimental techniques applied for the generation of heterogenous data sources considered in this study. On the one hand, experimental approaches are described which have been successfully applied for the analysis of genetic and epigenetic characteristics, including chromatin-immunoprecipitation (ChIP) and methylated DNA immunoprecipitation (MeDIP). On the other hand, concepts of DNA arrays and high-throughput sequencing are described. These technologies allow for genome wide analysis of experiment specific treated samples. In addition, I introduce statistical methods commonly used for analysing high-throughput data.

# 1 Introduction

In chapter 3, I present novel methods for analyzing MeDIP-seq data. As various computational methods already exist for analyzing gene expression, ChIP-on-Chip, and ChIP-Seq data [45, 17, 105], the major focus of this book is on the development of suitable methods for analyzing MeDIP-seq data. In detail, this book addresses not yet well researched issues like the development of novel quality control metrics for MeDIP-seq data (see section 3.2). These quality control metrics include estimation of the required sequencing-depth for obtaining reproducible full genome methylation profiles (*saturation analysis*, see subsection 3.2.1). As a second quality control metric, coverage of genome wide methylated CpG dinucleotides is presented (*coverage analysis*, see subsection 3.2.2). Third, a method for testing CpG enrichments within the short reads obtained by DNA sequencing is presented (*CpG enrichment analysis*, see subsection 3.2.3).

Most importantly, in section 3.3, I present a novel method for normalization of MeDIP-seq signals with respect to genomic regions of varying CpG densities. First of all, the transformation of raw MeDIP-seq data into a genome wide reads per million format is explained in section 3.3.1. Afterwards, I introduce the concept of coupling factors as a measure for local CpG densities (see section 3.3.2). In section 3.3.3, raw MeDIP-seq signals are connected to previously generated coupling factors and the dependency of local CpG densities and MeDIP-seq signals is revealed and visualized by calibration plots. Subsequently, I present the concept of normalization of raw MeDIP-seq signals into relative methylation scores (see section 3.3.4) and into absolute methylation scores (see section 3.3.5), respectively. In section 3.3.6, an evaluation of the developed normalization method is given. Here, improvements of normalization are shown by a comparison of raw and normalized methylation profiles against bisulphite derived methylation data. In addition, normalization results are compared to results of the only abvailable but computational demanding alternative MeDIP-seq data normalization approach [27]. In order to provide another important functionality for higher level data analysis, I have developed a statistical method for the identification of differentially methylated regions based on normalized MeDIP-seq data (see section 3.4).

I have implemented the several developed methods for immediate use as a standalone software package. Therefore, I present the first standard pipeline for the analysis of MeDIP-seq data. The entire computational approach (MEDIPS), including data processing, quality control, normalization, statistical analysis of differential methylation and methods for simulation of read coverage and saturation has been made available as an R software library (see section 3.5). MEDIPS is suitable for any arbitrary genome available via Bioconductor's annotation libraries [33] and more than 2 billion sequencing reads from mouse and human have been already processed.

In chapter 4, I apply the computational analysis approach to the analysis of cellular differentiation of human embryonic stem cells (hESCs). As hESCs are pluripotent, and therefore, can be induced to differentiate into a wide variety of cell types, these cells hold promise for cell replacement therapy [4]. Pluripotency and self renewal are controlled by a transcription regulatory network governed by the transcription factors OCT4, SOX2 and NANOG [13]. Recent studies on reprogramming somatic cells to induced pluripotent stem cells highlight OCT4 as a key regulator of pluripotency [96, 104]. In section 4.1, I present an integrated analysis of available high-throughput data, including ChIP-on-chip and RNAi experiments along with promoter sequence analysis of putative target genes. By this integrated approach, I have identified a core OCT4 regulatory network in human embryonic stem cells consisting of 33 target genes. [17].

Differentiation of hESCs along the endodermal lineage is induced by treatment with Activin A, a member of the TGF$\beta$ family of ligands [24, 3], resulting in definitive endoderm (DE). We derived DE cells from hESCs and analyzed the resulting transcriptome and methylome profiles of both cell types using the Illumina beadarray platform and MeDIP-seq technologies. The results of the quality control metrics for the obtained MeDIP-seq data from the two conditions and from additional sequenced input samples are presented in section 4.3.

The impact of normalization of generated MeDIP-seq data from hESCs is demonstrated by a comparison to recently published base-specific whole genome shotgun bisulphite sequencing (WGSBS) data of hESCs [64] (see section 4.4). The obtained MeDIP-seq methylation profiles are used for testing methylation patterns of promoter regions (see

# 1 Introduction

section 4.5) and of transcription factor binding sites (see section 4.6). By accessing normalized MeDIP-seq data from hESCs, DE, and with respect to the additional input data, genome wide differentially methylated regions (DMRs) are identified (see subsection 4.7). Analogous to Lister et al. [64], I identified a large number of de-methylation events (for example in the OCT4 transcription factor promoter), emphasizing an important role of de-methylation during differentiation of hESCs. Moreover, differential methylation was examined for further functional known genomic regions like promoters, exons, introns, and CpG islands (see section 4.8) and for previously identified transcription factor binding sites (see section 4.9).

Identified DMRs are further analyzed by an overrepresentation approach [11], connecting differential methylation to transcription factor binding, histone modifications, and to further functional known genomic attributes (see section 4.10). In order to test for the effect of de- and *de-novo* methylation on gene expression alterations, according hybridization experiments of hESCs and DE were performed and differential gene expression was calculated (see section 4.11). Finally, the effects of differential methylation and gene expression on the previously identified core regulatory OCT4 network that controls pluripotency is evaluated in detail in section 4.12.

In summary, this study provides novel concepts in the context of MeDIP-seq data analysis, including MeDIP-seq data normalization, quality control metrics, and identification of differential methylation. By applying MEDIPS to novel MeDIP-seq data, this study extends the knowledge on genome-wide regulatory modules and the interplay of genetic and epigenetic mechanisms during early endodermal differentiation of human embryonic stem cells. The results show that MEDIPS is an efficient approach for genome-wide methylation analysis that significantly reduces the imbalance of sequencing data generation and analysis and will assist further studies aiming to understand and characterize the function of DNA-methylation.

# 2 Experimental techniques

Genome wide genetic and epigenetic dependencies can be analysed by several different experimental approaches combined with high-throughput technologies. For example, DNA microarrays allow for analysing gene expression patterns of all known genes in parallel. When combined with appropriate experimental approaches, another recent format of DNA microarrays can be utilized in order to identify transcription factor binding events, histone modifications, or DNA methylation on a genome wide level. High-throughput sequencing is an emerging technology that increasingly replaces DNA arrays. This is because novel sequencing technologies allow for massive sequencing of millions of experiment specific DNA fragments in parallel and costs are permanently decreasing. The advantage of sequencing is that available DNA fragments can be detected without sequence specific limitations introduced by DNA arrays. Both high-throughput technologies, DNA microarrays and sequencing, produce enormous amounts of data. True biological observations have to be extracted from the data and suitable multivariate statistical methods have to be applied in order to distinguish signals from noise.

Chapter 2 of this book introduces into techniques underlying DNA arrays and high-throughput sequencing. In addition, experimental approaches are described which have been applied in order to analyze genetic and epigenetic characteristics in human embryonic stem cells. Moreover, this chapter introduces basic statistical methods commonly applied for the analysis of high-throughput data. Experimental techniques described in this chapter were selected because they are source of the heterogenous data types which serve as the basic input for data analysis methods developed and applied in the context of this book.

## 2.1 DNA arrays

DNA microarrays are a widely used technology in molecular biology [92, 65, 66]. It is composed of a carrier substrate (e.g. coated plastic or coated glass) which contains immobilized DNA sequences (probes), specific for genetic regions of a reference organism. For example, microarrays are used for measuring the gene expression of thousands of genes in parallel. Here, the probes are designed for being complementary to selected transcribed regions of genes, typically in the surrounding of the 3'-end. The mRNA of the targeted biological material is extracted and marked (e.g. radioactively or fluorescently). With a hybridization experiment, the marked cRNA is attached to complementary probes on the microarray and the strength of the resulting probe-specific signals is considered as an indicator for the expression of the according genes within the targeted material. Microarrays differ in their surface material, the technique of immobilizing the probes and in the way of labeling of taget material.

A recent microarray format is the Illumina BeadChip system [52]. The *HumanRef-8* expression bead chip contains >24,000 unique sequences immobilized on magnetic beads which generate the array elements. Each bead contains hundreds of thousands of copies of covalently attached oligonucleotide probes. These beads are quantitatively pooled and introduced to etched microwell substrates. Once introduced, the beads spontaneously assemble into wells, resulting in a high density microarray. After bead assembly, a hybridization-based procedure is used to map the array, determining which bead type resides in each well. Figure 2.1 illustrates one oligomer attached to a bead. Each oligonucleotide is of length 75nt comprising an 25nt identifier sequence (address) and a 50nt gene-specific probe. The identifier sequence is used for identifying the location of the different bead types on the microarray. The figure shows a biotin labeled cRNA from a target sample hybridized to the gene-specific probe.

An important application of gene expression microarrays is the identification of differentially expressed genes, i.e. those genes which show a significant alteration of their expression when testing two biological samples of different origins. In this book (see chapter 4), the gene expression profile of undifferentiated hESCs (control) is compared to the gene expression profile of differentiated definitive endoderm (treatment). Dependent on

## 2 Experimental techniques

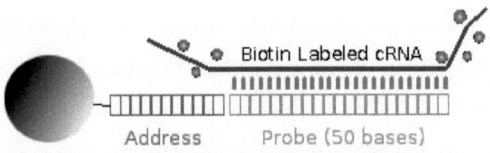

**Figure 2.1: BeadChip technology.** Gene specific probes are attached to beads which are then assembled into the arrays [52]. Here, only one oligomer is attached to the bead; actual beads have hundreds of thousands of copies of the same sequence attached. Each oligonucleotide is of length 75nt comprising an 25nt identifier sequence (address) and a 50nt gene-specific probe. The identifier sequence is used for identifying the location of different bead types on the microarray. A biotin labeled cRNA from the target sample hybridized to the gene-specific probe is shown in a schematic way.

the type of microarray, both samples are either labeled in the same way and are subsequently subjected to separated microarrays (this is for oligonucleotide arrays) or both samples are labeled using two distinct fluorescent dyes and are subsequently subjected to the same microarray (two-color experiment for cDNA arrays). In both approaches, gene expression for both biological samples is deduced by the (fluorescent specific) signal intensities of the gene specific probes. For a statistical analysis of differential gene expression, such microarray experiments have to be performed in a sufficient number of biological replicates. In order to compare gene expression profiles derived from several microarray experiments, the obtained intensity signals have to be normalized due to differing signal intensity ranges caused by variable fluorescence labeling, material availability, and hybridization efficiency. Several methods for microarray normalization have been developed [61, 29, 105].

Based on the normalized signal intensities, differential gene expression is calculated separately for each gene. For this, the average of gene-associated signal intensities of the control sample is divided by the average of the according gene-associated signal intensities of the treatment sample. Complementary to this fold-enrichment, statistical tests are applied in order to test, if the distribution of gene specific signal intensities of the control group differs significantly from the distribution of the according gene specific signal intensities of the treatment group (see section 2.2). The combination of fold-enrichment

## 2 Experimental techniques

and statistical testing has proven to be a practical and sophisticated approach for identification of differential gene expression.

Another type of microarrays are tiling arrays. In contrast to gene expression arrays, tiling arrays are used for identifying genome wide DNA fragments. For this application, oligonucleotides are generated which are complementary to continuous genomic regions of a reference genome (see Figure 2.2 B). The relation between length and distance inbetween those probes define the density of the tiling array and therefore, also the accuracy of the subsequent allocation of detected DNA fragments. Such probes typically cover genomic regions of interest, e.g. the promoter regions of genes or even the entire genome of interest. For the latter case, tiling arrays typically consist of a set of complementary arrays.

In the context of tiling arrays, the targeted biological material is treated in another way than for gene expression arrays. Here, not the mRNA is extracted but the DNA is sheared into small DNA fragments. This can be achieved by sonicating the DNA material and typically results in fragments of 300-700 bp length. The full amount of DNA fragments serves as control (Input) sample. In a separate procedure, these DNA fragments are enriched, typically by immunoprecipitation, for a specific characteristic to be analyzed. This can be either local DNA-protein binding events (see Figure 2.2 A and section 2.4) or methylated cytosines (see section 2.5). The experiment specific enriched DNA fragments serve as the treatment (immunoprecipitated, IP) sample. As for gene expression experiments, the Input and IP samples are labeled (either as for one- or for two-colour experiments) and are subsequently subjected to the tiling array(s) (see Figure 2.2 A). For each oligonucleotide, the absolute IP and Input intensity values are stored and can be compared to each other (see Figure 2.2 C).

The identification of statistical significant differences between the enriched and nonenriched samples, in principle, follows the statistical analysis used for gene expression microarrays (see section 2.2) but tiling array specific algorithms have been developed [17, 45, 46]. Such methods make use of the circumstance that neighboring probes (relative to their location on the reference genome) will simultaneously detect enriched DNA fragments and their resulting signal intensities will follow a peak-like distribution (see Figures 2.2 B and D). The array-specific density of tiled probes and the estimated average length of immunoprecipitated DNA fragments define the number of neighboring probes that will give rise to local signal enrichments and therefore, define the shape of the resulting peaks. Tiling array specific computational methods incorporate these information for the identification and localization of enriched DNA fragments.

## 2.2 Statistical testing of differential expression

Statistical tests have proven to be a sensitive approach for identifying differentially expressed genes on the basis of microarray experiments [21, 40, 38]. In the context of this book, statistical tests are also applied for identifying differentially methylated regions on the basis of pre-processed MeDIP-seq data (see section 3.4). The results of such tests are generally combined with further calculated attributes in order to exclude less distinctive absolute differences of the two compared data distributions. This can be either a threshold of minimal required signal intensity derived from background data, or thresholds for the ratio between condition-wise averaged control and treatment signal intensities.

For statistical testing, a sufficient number of experimental (e.g. microarray) repetitions have to be performed. By a repetition of any such measurements of two different conditions, for each tested attribute (e.g. for each probe), there will be two samples of intensity values $X_1,...,X_N$ and $Y_1,...,Y_M$ (control and treatment). Statistical modelling requires the two samples to conform to a certain probability distribution, e.g. a normal distribution. Two hypotheses are then composed:

$H_0$: The samples have the same location (null hypothesis)
$H_1$: The samples have different locations (alternative hypothesis)

Statistical tests allow for rating, if the two samples are derived from the same population (no differential expression) or not (differential expression). Statistical tests applied in the context of this book are

- Student's t-test with equal variances, and

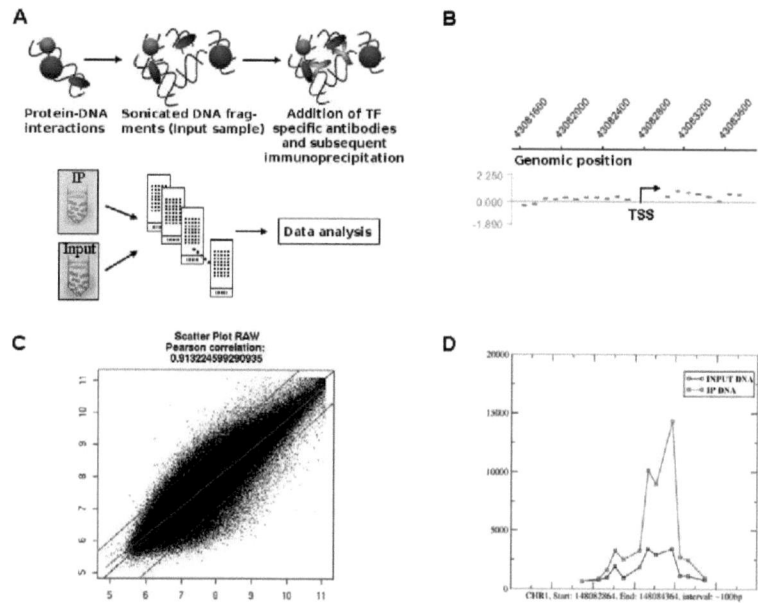

**Figure 2.2: Chromatin immunoprecipitation (ChIP). A** Protein-DNA interactions are fixed by adding formaldehyde to the targeted biological sample. By sonication, the DNA is sheared into fragments of length 0.2-1kb. Selected DNA-protein complexes are enriched (immunoprecipitated) by adding TF specific antibodies (IP sample). As a control, a fraction of the sheared DNA is treated with an unspecific antiobody and the resulting sample is typically called the genomic or input sample. Subsequently, a series of two-colour tiling array experiments are performed. **B** Genomic regions of interest (here the promoter region around the transcription start site (TSS) of a gene) are continuously covered by complementary oligonucleotides. **C** Comparison of absolute intensity values from the IP- and input samples (log2). **D** Example of a chromosomal region bound by a specific antibody. The intensity values of neighboring oligonucleotides are increased by a given factor within the IP sample (red curve) compared to the Input sample (black curve).

## 2 Experimental techniques

- Wilcoxon's rank sum test (also Mann-Whitney U test)

Based on a test-statistic, i.e. a mathematical function applied to the data, both tests calculate a p-value allowing for rating the significance of differential expression. The test-statistic of the the t-test is e.g.

$$T(X_1,...,X_N,Y_1,...,Y_M) = \frac{\overline{X}-\overline{Y}}{\sqrt{\frac{(N-1)S_X^2+(M-1)S_Y^2}{N+M-2}}}\sqrt{\frac{N \cdot M}{N+M}} \qquad (2.1)$$

where $S_X^2$ and $S_Y^2$ are the empirical variances of the control and the treatment group, i.e.

$$S_X^2 = \frac{1}{N-1}\sum_{i=1}^{N}(X_i-\overline{X})^2 \qquad (2.2)$$

$$S_Y^2 = \frac{1}{M-1}\sum_{i=1}^{M}(Y_i-\overline{Y})^2 \qquad (2.3)$$

In addition, $\overline{X}$ and $\overline{Y}$ are the according averages:

$$\overline{X} = \frac{1}{N}\sum_{i=1}^{N}X_i \qquad (2.4)$$

$$\overline{Y} = \frac{1}{M}\sum_{i=1}^{M}Y_i \qquad (2.5)$$

This test statistic follows a determined distribution. Given that the control and treatment samples are normal distributed, this is a t-distribution with $M+N-2$ degrees of freedom. For each experimental result, a p-value can be calculated that indicates the probability for the t-distribution to result in a more extreme value than the observed one. Therefore, a small p-value indicates an event of differential expression. The p-value itself is considered as a significance measure for the deviation of the data from the null hypothesis.

T-tests are parametric approaches that require the data to emerge from a probability distribution which can be described by a mathematical function (e.g. a normal distribution). Alternatively, Wilcoxon's rank sum test is a non-parametric approach that allows the data to emerge from less stringent defined distributions. It can therefore be applied to an extended class of problems and is especially suitable in cases where it cannot be ensured that measured signal intensities are normal distributed. For each tested attribute

## 2 Experimental techniques

(e.g. for each probe), the Wilcoxon's rank sum test again considers the repetitive and independent (microarray) measurements $X_1,...,X_N$ and $Y_1,...,Y_M$ of two different conditions (control and treatment). The main assumption is that both random variables follow continuous distribution functions $F_1$ and $F_2$ and differ only by a shift of $\delta$:

$$F_1(x) = F_2(x - \delta) \tag{2.6}$$

The test involves the calculation of an U statistic whose distribution under the null hypothesis is known. In case of small samples, the distribution can be calculated, but for larger sample sizes its distribution is approximated by the normal distribution. For each tested attribute, the U test starts by merging all observations from the control and treatment experiments into one single ranked series. In case there are two or more identical values in both series, these values are first ranked in an arbitrary order but subsequently their ranks are replaced by their combined rank average. Based on this merged series, for each group, the sum of all ranks are determined as

$$R_1 = \sum_{i=1}^{N} R(X_i) \tag{2.7}$$

$$R_2 = \sum_{i=1}^{M} R(Y_i) \tag{2.8}$$

where $R(X_i)$ and $R(Y_i)$ are the ranks of the i-th measurement of the control or the treatment sample, respectively. In fact, the sum of all ranks equals

$$R_1 + R_2 = \frac{(N+M)(N+M+1)}{2} \tag{2.9}$$

Therefore, after having calculated the sum of all ranks for one group (e.g. $R_1$), the other one follows by calculation. For both groups, individual U values are calculated as

$$U_1 = R_1 - \frac{N(N+1)}{2} \tag{2.10}$$

$$U_2 = R_2 - \frac{M(M+1)}{2} \tag{2.11}$$

Subsequently, it is valid

## 2 Experimental techniques

$$U_1 + U_2 = N \cdot M \tag{2.12}$$

This is because from formula 2.9, it follows:

$$U_1 + U_2 = R_1 - \frac{N(N+1)}{2} + R_2 - \frac{M(M+1)}{2}$$

$$= \frac{(N+M)(N+M+1)}{2} - \frac{N(N+1)}{2} - \frac{M(M+1)}{2} = N \cdot M$$

For the subsequent test, the minimal of both U values is considered:

$$U = min(U_1, U_2) \tag{2.13}$$

As mentioned above, for larger sample sizes, the U distribution can be approximated by the normal distribution. In that case, the standardized value

$$Z = \frac{U - \mu_U}{\sigma_U} \approx \mathcal{N}(\mu = 0; \sigma^2 = 1) \tag{2.14}$$

is a standard normal variable, where

$$\mu_U = \frac{N \cdot M}{2} \tag{2.15}$$

is the mean and

$$\sigma_U = \sqrt{\frac{(N \cdot M)(N + M + 1)}{12}} \tag{2.16}$$

is the standard deviation of $U$. The levels of significance result by the levels of significance of the approximated standard normal distribution $\mathcal{N}(\mu = 0; \sigma^2 = 1)$ and therefore, the significance of the calculated $Z$ value can be deduced by the standard normal distribution. The according p-value indicates the probability for the U-distribution to result in a more extreme value than the observed one. Therefore, a small p-value indicates an event of differential expression. As an example, by defining a level of significance to a concrete percentage, e.g. $p = 0.01$, for the outcome of the alternative hypothesis (differential

## 2 Experimental techniques

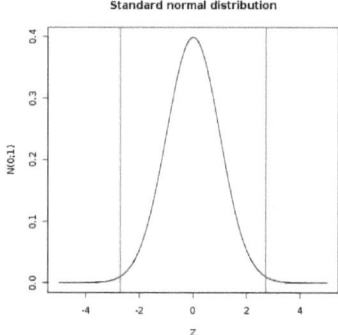

**Figure 2.3: Standard normal distribution.** Because the U distribution is approximately normally distributed, for the standard normal variable $Z$ it is valid: $Z \approx \mathcal{N}(\mu = 0; \sigma^2 = 1)$. The figure shows the range of the normal distribution for $Z$ values ranging from -5 to 5. Here, the significance level for $p = 0.01$ are indicated as red lines at both sides of the distribution. The according $Z$ values are -2.715 and +2.715. For rejecting the null hypothesis (differential expression) on that significance level, the data derived $Z$ value has to be outside of these extreme values.

expression) in a two-tailed test, it must be valid (see Figure 2.3):

$$Z \notin [-2.715, +2.715]$$

By this approach, the Wilcoxon's rank sum test allows to rate whether one of two measured data series is significantly larger than the other.

## 2.3 Second generation sequencing

DNA sequencing is the process of determining the order of the nucleotide bases (adenine, guanine, cytosine, and thymine) in a molecule of DNA. Sequencing of the full human genome was achieved by a technique developed by Frederick Sanger [90].

The classical Sanger sequencing (also called chain-termination method) requires a single-stranded DNA template, a DNA primer, a DNA polymerase, radioactively or fluorescent labeled nucleotides, and modified nucleotides that terminate DNA strand elongation [90]. The DNA sample is divided into four separate sequencing reactions, containing all four of the standard deoxynucleotides and the DNA polymerase. To each reaction is added only one of four di-deoxynucleotides which are the chain-terminating nucleotides, thus terminating DNA strand extension and resulting in DNA fragments of varying length [90, 91]. The newly synthesized and labeled DNA fragments are heat denatured, and separated by size with a resolution of just one nucleotide by gel electrophoresis. The DNA bands are then visualized by UV light, and the DNA sequence can be directly read off the gel image [90, 91, 69].

The high demand for low-cost sequencing has driven the development of high-throughput sequencing technologies that parallelize sequencing processes, producing thousands or millions of sequences at once [35, 19]. High-throughput sequencing technologies are intended to lower the cost of DNA sequencing beyond what is possible with standard dye-terminator methods [94]. A novel high-throughput sequencing method uses bridge PCR for *in vitro* clonal amplification, where fragments are amplified upon primers attached to a solid surface. This technology is used by the Illumina Genome Analyzer (www.illumina.com). For high-throughput sequencing, DNA molecules are physically bound to a surface, and sequenced in parallel. This technology allows massive parallel sequencing of millions of fragments using a reversible terminator-based sequencing chemistry.

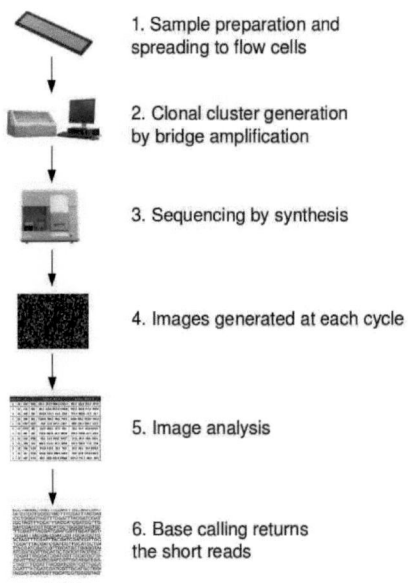

Figure 2.4: **Process of Illumina sequencing.** The workflow is described in the adjacent text. (Individual images were taken from http://www.illumina.com.)

Alternative sample preparation methods allow the sequencing systems to be used for a range of applications including gene expression, ChIP, and MeDIP. After having performed the experiment of interest on the targeted biological material, libraries have to be generated by adapter ligation to the DNA fragments before spread on the flow cells (step 1 in Figure 2.4). This step is necessary for the subsequent cluster amplification, as Illumina's sequencing technology relies on the attachment of randomly fragmented genomic DNA to a planar, optically transparent surface. Attached DNA fragments are extended and bridge amplified to create high density sequencing flow cells with hundreds of millions of clusters, each containing $\sim$1,000 copies of the same template (step 2). These templates are sequenced using a four-color DNA sequencing-by-synthesis technology that employs reversible terminators with removable fluorescent dyes. The sequencing-by-synthesis approach typically runs in 36 cycles (step 3). After each cycle, an image is generated by high-sensitivity fluorescence detection using laser excitation and total internal reflection optics (step 4). The first step in the primary analysis is interpreting the image data in order to identify distinct clusters and to create digital intensity files describing the signal intensities of each cluster in each cycle (step 5). Signal intensity profiles for each cluster are used to call bases. Determining the quality of each base call is crucial for downstream analysis and confidence scores for each call are calculated (step 6). Finally, millions of quality tagged short reads are received, typically of length 36bp. The obtained sequence reads are aligned against the according reference genome and finally application specific data analysis tools are applied.

Image analysis, base calling, and efficient quality score dependent alignments are currently intensively explored topics in the field of computational biology. However, these topics are not subject of this book. Proprietary software for the image analysis and base calling as well as an available sophisticated alignment software [57] are applied. The main focus of this book is on the development of novel methods for the analysis of MeDIP specific sequencing data. Especially a method is developed that corrects for a CpG density dependent bias which occurs due to preparation procedures during MeDIP experiments.

## 2.4 Chromatin immunoprecipitation (ChIP)

Chromatin immunoprecipitation (ChIP) is an experimental technique used to investigate interaction events between transcription factors (TFs) and DNA. In order to regulate gene expression, TFs bind to their sequence specific DNA binding sites across the genome, often but not exclusively in promoter regions of genes. Moreover, ChIP is also used for the identification of genomic locations associated to histone modifications [23].

The first step of a ChIP experiment is to stabilize protein-DNA interactions by adding formaldehyde [76]. Subsequently, DNA is sheared by sonication resulting in DNA fragments of length 0.2-1kb. Afterwards, the sample contains both protein bound and unbound DNA fragments. By adding a TF specific antibody, enrichment of such DNA fragments bound by the according TF is initiated [89]. The treated sample is typically indicated as the IP (immuno-precipitated) sample. From the experimental point of view, the availability or generation of a highly TF specific antibody is most important. In each ChIP approach, the TF binding events of only one TF is examined. Increased enrichment of TF specific DNA fragments is achieved by PCR amplification. As a control, a fraction of the sheared DNA is treated with an unspecific antibody and the resulting sample is typically called the genomic or input sample (see Figure 2.5).

Transcription factor binding events can finally be deduced by identification of genomic regions that show enriched levels of associated DNA fragments in the IP sample compared to the Input sample. For this, it is necessary to detect and quantify the DNA fragment concentration along all chromosomes in both samples. DNA fragments can be either detected by tiling arrays (ChIP-on-Chip, see also section 2.1 and Figure 2.2 B-D) or by second generation sequencing (ChIP-Seq, see also section 2.3 and Figure 2.5). Several methods for determining statistically significant enriched genomic regions from ChIP experiments are available for ChIP-Chip [59, 99, 17, 46] and ChIP-Seq data [45, 101, 67, 14, 88].

## 2.5 Methylated DNA immunoprecipitation (MeDIP)

Methylated DNA immunoprecipitation (MeDIP) uses an antibody specific for methylated cytosines in order to immunocapture methylated genomic fragments [103]. First, genomic DNA is extracted from cells. Then purified DNA is subjected to sonication in order to shear it into random fragments (see Figure 2.6). The short length of these fragments is important for obtaining adequate resolution, improving the efficiency of downstream steps

## 2 Experimental techniques

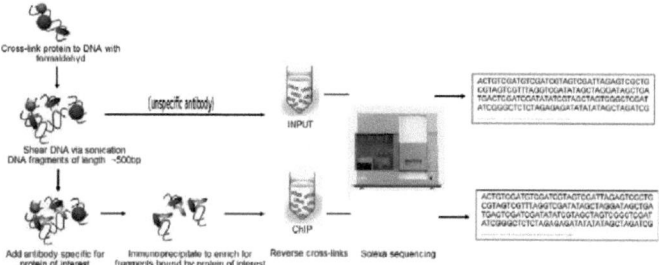

**Figure 2.5: Chromatin immunoprecipitation followed by sequencing (ChIP-Seq).** Protein-DNA interactions are cross-linked and stabilized by adding formaldehyde. Subsequently, DNA is sheared by sonication resulting in DNA fragments of length 0.2-1kb. Enrichment of bound DNA fragments is achieved by precipitating with a TF specific antibody (ChIP sample). As a control, a fraction of the sheared DNA is treated with an unspecific antibody and the resulting sample is typically called the genomic or input sample. Transcription factor binding events can be deduced by identification of genomic regions that show enriched levels of DNA fragments in the ChIP- compared to the input sample. For this, it is necessary to detect and quantify the DNA fragment concentration along all chromosomes in both samples. DNA fragments can be detected by second generation sequencing (ChIP-Seq). (The image of the sequencer was taken from http://www.illumina.com.)

in immunoprecipitation, and reducing fragment-length effects or biases [70, 42]. To further improve binding affinity of antibodies, the DNA fragments are denatured to produce single-stranded DNA. Following denaturation, the DNA is incubated with monoclonal antibodies which bind to methylated cytosines. Subsequently, classical immunoprecipitation technique is applied (see section 2.4). DNA is purified from antibodies by enzymatic digestion and afterwardds prepared for DNA detection [83, 106, 108]. Analogous to ChIP (see section 2.4), immunoprecipitated methylated DNA fragments can be detected either by tiling arrays (MeDIP-Chip) or by next-generation sequencing (MeDIP-seq).

As already mentioned in section 1.4, it has been shown that MeDIP derived data needs to be corrected for local CpG densities in order to estimate valid methylation levels [27, 80]. This effect is caused by a varying efficiency of antibody binding and immunoprecipitation dependent on the local density of methylated CpG sites (see Figure 2.6). Especially the analysis of mCpG-poor regions has been assumed to be difficult [103, 27]. While there is applicable software available for analyzing MeDIP-Chip data [27, 80], normalization of MeDIP-seq data is in principle solved [27] but remains disproportional time-consuming.

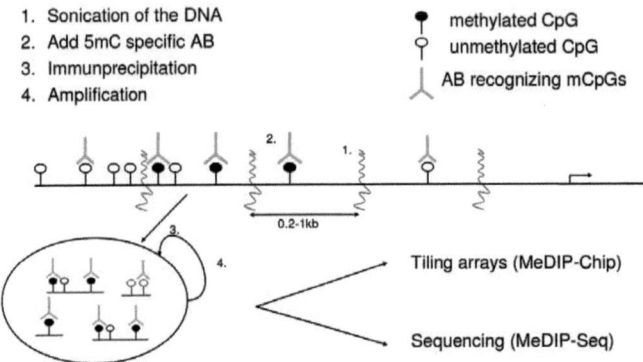

**Figure 2.6: MeDIP workflow overview.** In principle, the workflow follows classical ChIP approaches (see section 2.4) but here, an antibody specific for methylated cytosines (5mC) is utilized. The figure illustrates the varying efficiency of antibody binding and immunoprecipitation dependent on the local density of methylated CpG sites. Especially the analysis of mCpG-poor genomic regions has to be considered critical [27]. MeDIP is followed either by array-hybridization (see also section 2.1) or by second generation sequencing (see also section 2.3).

## 2.6 Alternative methods for detecting DNA methylation

Although the main focus of this book is on the development of methods for analyzing MeDIP-seq data, there are several further methods available for detecting DNA methylation. The impact of the developed MeDIP-seq data normalization method (see section 3.3) is evaluated by comparing the resulting methylation profiles to methylation profiles derived by other techniques (see subsection 3.3.6). Therefore, an overview of such alternative techniques is given in this section.

The discovery that treatment of denatured genomic DNA with sodium bisulphite chemically deaminates unmethylated cytosine residues much more rapidly than methylated cytosines [37, 102] spurred a revolution in DNA methylation analysis in the 1990s [31, 20, 54]. This chemical treatment of DNA effectively turns an epigenetic difference into a genetic difference — unmethylated Cs are converted to Ts (by uracil) — thereby enabling many new DNA methylation detection and analysis techniques [53, 83, 8, 100, 36, 106, 62, 60, 42, 9, 44]. Analysis of bisulphite-converted DNA was initially done by Sanger sequencing

## 2 Experimental techniques

of cloned PCR products from single loci [31, 79]. Many improvements have since then been developed, including quantitative direct Sanger sequencing of PCR products [20] and highly automated application of this approach [30, 54]. By the latter approach, base specific methylation values were generated for selected genomic regions on the chromosomes 6, 20, and 22 in several human tissues, including e.g. sperm samples [30]. In subsection 3.3.6, these bisulphite derived sperm methylation patterns are compared against unnormalized and normalized MeDIP-seq data derived from another human sperm sample [27].

Bisulphite genomic sequencing excels at producing base-pair resolution DNA methylation information, but bisulphite-based methods are not easily adapted to array-hybridization techniques and so, until recently, were rarely used for genome-scale DNA methylation analysis [54]. With the exception of 5-Methylcytosine (5mC) residues, bisulphite-treated DNA is comprised of three different bases instead of four. This reduced sequence complexity, and therefore greater sequence redundancy, results in decreased hybridization specificity [54]. Hybridization of bisulphite-converted DNA either requires dedicated tiling arrays designed for the bisulphite-converted genome or must allow for substantial mismatches in hybridization [54].

Although the adaptation of bisulphite-converted DNA to array hybridization has been challenging, bisulphite-converted DNA is particularly well suited for sequencing-based approaches and is now enjoying a resurgence thanks to the application of next-generation sequencing platforms [54]. In order to restrict the full amount of DNA that has to be subsequently analyzed, restriction endonucleases can be utilized. The most widely used methylation-sensitive restriction enzymes for DNA methylation studies are HpaII and SmaI. Reduced representation bisulphite sequencing (RRBS) was introduced to reduce sequence redundancy by selecting only some regions of the genome for sequencing by size-fractionation of DNA fragments after BglII digestion [71] or after MspI digestion [72]. These choices of restriction enzymes enrich for CpG-containing segments of the genome but do not target specific regions of interest in the genome [54].

Restriction enzyme enrichment techniques are currently being adapted so that the readout can be obtained by next-generation sequencing techniques instead of array hybridization. Sequence-based analysis is more flexible and powerful as it allows for allele-specific DNA methylation analysis, does not require an appropriately designed microarray, can cover more of the genome with less input DNA and avoids hybridization artefacts, although it is still subject to sequence library biases [54]. Sequencing-by-synthesis of libraries constructed from size-fractionated HpaII or MspI digests that are compared with randomly sheared fragments is known as Methyl–seq [15].

The most comprehensive single-base-pair resolution DNA methylation analysis technique is whole-genome shotgun bisulphite sequencing [54]. Whole-genome shotgun bisulphite sequencing (WGSBS) has been achieved on the Illumina Genome Analyzer platform for small eukaryotic genomes, such as *Arabidopsis thaliana* [63, 22], and for mammalian DNA [64]. Increased read lengths and paired-end sequencing strategies have aided the implementation of WGSBS [28, 25, 51], although approximately a tenth of the CpG dinucleotides in the mammalian genome remain refractory to alignment of bisulphite-converted reads [54]. In chapter 4, available WGSBS data from Lister et al. [64] is compared to novel MeDIP-seq data generated in the context of this book. It is demonstrated that correlation of WGSBS and MeDIP-seq data is significantly improved after normalization of MeDIP-seq data. The developed procedure of MeDIP-seq data normalization is described in chapter 3.

# 3 Modelling of MeDIP-seq data

The major focus of this book is on the development of computational methods suitable for analyzing MeDIP-seq data. In chapter 3, I present novel methods for the analysis of MeDIP-seq data, including time-efficient raw data processing (see section 3.1), quality control metrics (see section 3.2), normalization (see section 3.3), and identification of differential methylation (see section 3.4). In section 3.3.6, improvements of normalization are shown by a comparison of raw and normalized methylation profiles against bisulphite derived methylation data. In addition, normalization results are compared to results of the only abvailable but computational demanding alternative MeDIP-seq data normalization approach [27]. The entire computational approach (MEDIPS), including data processing, quality control, normalization, statistical analysis of differential methylation and methods for simulation of read coverage and saturation has been made available as an R software library. In section 3.5, I present this first standard pipeline for the analysis of MeDIP-seq data.

## 3.1 Genome vector

Methylated DNA immunoprecipitation (see section 2.5) followed by sequencing (see section 2.3) results in millions of experiment specific short DNA sequences. In order to identify their genomic origin, they are aligned to the according reference genome by applying available alignment implementations like e.g. MAQ [57] or Bowtie [55]. Furthermore, standard post-processing of the alignment results is to filter out mapped reads of low quality and to exclude artificial short read pile-ups. For all remaining short reads, their genomic coordinates together with their associated strand information (plus or minus strand) are extracted and serve as the basic information obtained from a MeDIP-seq experiment.

3 Modelling of MeDIP-seq data

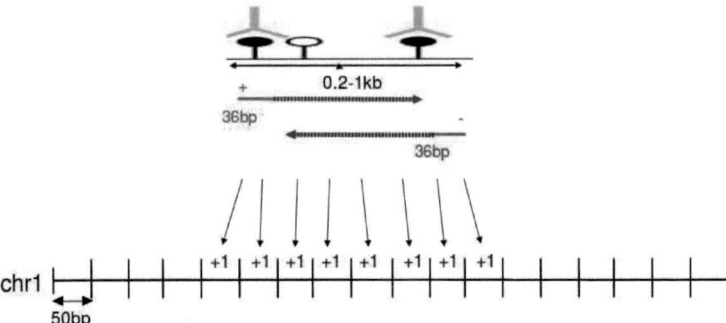

**Figure 3.1: The genome vector.** MeDIP-seq derived short reads (red lines) are mapped against the according reference genome. Based on their obtained genomic coordinates, they are extended to a length of e.g. 400bp along the plus or the minus strand according to the associated strand information (blue dashed lines). Genome wide short read coverage is calculated by first defining a targeted resolution (here 50bp bins) and by counting the number of overlapping extended reads at each genomic bin position. The genome vector is the computational representation of the short read coverage at a 50bp bin resolution (single chromosome vectors are concatenated to one genome vector).

In order to calculate the genome-wide short read coverage, a targeted data resolution has to be determined. In principle, a short read coverage can be calculated for each base position. Because the resolution of MeDIP-seq data is restricted by the size of the sonicated DNA fragments after amplification and size selection (typically between 0.2-1kb), a bin size of 50bp is considered as a reasonable compromise on data resolution and computational costs. Moreover, short reads generated by modern-day sequencers do not represent the full DNA fragments but are of shorter length (e.g. 36bp). Therefore, the data is smoothed by extending each read to a length according to the estimated average length of sequenced DNA fragments, either along the plus or along the minus strand, as specified by the short read dependent strand information. Each chromosome is then divided into bins of size 50bp and the short read coverage is calculated on this resolution. In the following, the bin representation of the genome is called the *genome vector* (see Figure 3.1)

Second generation sequencing approaches generate millions of short reads per experiment. Long-term data storage and efficient data processing are challenging tasks, even for modern-day servers. Naive programming approaches cannot be applied in appropriate time. For example, the task of identifying overlapping extended short reads at genome wide 50bp bins needs to be implemented in a sophisticated way. Hence, most of the algorithms implemented in the context of this work are primary runtime optimized.

Algorithm 3.1 shows the R source code as implemented within the MEDIPS package (see section 3.5) for calculating the short read coverage. In fact, this function is called by a wrapper function which declares and initializes the genome vector with respect to the number and lengths of chromosomes of the reference genome and with respect to the targeted coverage resolution (e.g. 50bp). The given *MEDIPS.distributeReads* function is called from a loop processing the chromosomes. The wrapper function provides separated vectors for the start, stop, and strand informations of the reads (all of the same length $n$, where $n$ is the number of short reads available for the current chromosome), as well as a vector containing the chromosomal positions of the predefined bins within the current chromosome. The latter vector is of length $m$, defined by simply dividing the length of the current chromosome by the targeted resolution (e.g. $binSize$=50bp)

$$m = \lfloor \frac{\text{length}(chromosome)}{binSize} \rfloor \qquad (3.1)$$

First of all, the short reads are extended to a length as specified by the *extend* parameter. Moreover, for each of the provided vectors, additional identification ($id$) vectors are generated. For the vector containing the start positions of the short reads, an $id$ vector of length $n$ is defined (here *reads_start_id*) and each entry is assigned to the value 1. For the vector containing the stop positions of the short reads, an $id$ vector of length $n$ is defined (here *reads_stop_id*) and each entry is assigned to the value -1. For the vector containing the chromosomal bin positions, an $id$ vector of length $m$ is defined (here *positions_id*) and each entry is assigned to the value 0.

In addition, two vectors *ct.vec_pos* and *ct.vec_id*, both of length $2n+m$, are declared. The start and stop positions of the short reads as well as the bin positions are concatenated and assigned to the *ct.vec_pos* vector. Additionally, the previously created $id$ vectors are concatenated in the same order as the position vectors and are assigned to the *ct.vec_id* vector. This combined $id$ vector is sorted with respect to the order of the sorted combined position vector (this is the sorted *ct.vec_pos* vector). From the algorithmic point of view, this sorting is the most time consuming step that can be solved in an average runtime of $\mathcal{O}((2n+m)\log(2n+m))$ (e.g. using quicksort).

Next, R's *cumsum()* function is applied to the ordered combined *id* vector (this is the ordered *ct.vec_id* vector). The cumulative sum is a sequence of partial sums of a given sequence. For example, given a sequence $\{x_k\}_{k=1}^{2n+m}$, a partial sum of the first $j$ terms is given by $count_j = \sum_{k=1}^{j} x_k$. For example, the cumulative sums of a sequence $(a, b, c, ...)$ are $a, a+b, a+b+c, ....$ Therefore, the *cumsum()* function returns a vector of length $2n+m$, where each entry $j$ results by the cumulative sum along the ordered combined *id* vector (this is the ordered *ct.vec_id* vector):

$$count(x_1, ..., x_{d-1}, j, x_{d+1}, ..., x_{2n+m}) = \sum_{k=1}^{j} ct.vec\_id(x_1, ..., x_{d-1}, k, x_{d+1}, ..., x_{2n+m})$$

It is obvious that this calculation can be achieved in $\mathcal{O}(2n+m)$ calculation steps. The resulting *count* vector is considered as a counter that starts at zero. The *cumsum()* function increases the counter by 1, whenever a short read starts and decreases the counter by 1, whenever a short read stops. Because the chromosomal bin positions are associated to zeros, the *cumsum()* function does not change the counter at that positions. The chromosomal bin positions were sorted in between all start and stop positions of the short reads, and therefore, the current value of the counter at the chromosomal bin positions reflects the number of 'open' or overlapping short reads. From the resulting *count* vector, the number of overlapping short reads at each chromosomal bin position can be directly sorted out by selecting all entries whose indices are associated to zeros in the combined and ordered *id* vector (here *ct.vec_id*).

By this implementation, the task of identifying all overlapping short reads at arbitrary chromosomal bin positions is limited by $\mathcal{O}((2n+m)\log(2n+m))$ in time. Naive approaches easily end up in $2n \cdot m$ calculation steps when comparing all start and end positions of the short reads to the chromosomal bin positions. When considering MeDIP-seq data, $n$ has to be at least $2 \cdot 10^7$ for a well covered human methylome (see subsection 3.2.1). As the human genome consists of approximately $3 \cdot 10^9$ base pairs, there will be $m = 6 \cdot 10^7$

## 3 Modelling of MeDIP-seq data

**Algorithm 3.1 Calculating short read coverages in** $\mathcal{O}((2n+m)log(2n+m))$. Here, $n$ is the number of available short reads and $m$ is the number of genome wide bins as defined by the specified *binSize* parameter and by the size of the reference genome.

```
MEDIPS.distributeReads<-function(reads_start=NULL, reads_stop=NULL,
    reads_strand=NULL, positions=NULL, extend=0){

    if(extend!=0){
        reads_start[reads_strand=="-"]=reads_stop[reads_strand=="-"]-
            extend
        reads_stop[reads_strand=="+"]=reads_start[reads_strand=="+"]+
            extend
    }

    reads_start_id=vector(length=length(reads_start), mode="numeric")
    reads_start_id[]=1
    reads_stop_id=vector(length=length(reads_stop), mode="numeric")
    reads_stop_id[]=-1
    positions_id=vector(length=length(positions), mode="numeric")
    positions_id[]=0

    ctmatrix_pos=vector(length=length(reads_start)+length(reads_stop)
        +length(positions),mode="numeric")
    ctmatrix_id=vector(length=length(reads_start)+length(reads_stop)+
        length(positions),mode="numeric")
    ctmatrix_pos[]=append(append(reads_start,reads_stop), positions)
    ctmatrix_id[]=append(append(reads_start_id, reads_stop_id),
        positions_id)

    ctmatrix_id=ctmatrix_id[order(ctmatrix_pos)]
    count=cumsum(ctmatrix_id)

    return(count[ctmatrix_id==0])
}
```

genomic bins when defining a bin size of 50bp. Therefore, the presented implementation is limited by $(2n+m)\log(2n+m) = (1 \cdot 10^8)\log(1 \cdot 10^8) = 1,842,068,074$ calculation steps, whereas the naive approach will need $2n \cdot m = 2 \cdot 2 \cdot 10^7 \cdot 6 \cdot 10^7 = 2,400,000,000,000,000$ calculation steps. This example emphasizes the need for time efficient implementations when high-throughput sequencing derived data is modelled.

The described method for calculating the genome vector is only one of several implementations within the MEDIPS package (see section 3.5) optimized for runtime. However, the following sections describe the developed concepts for MeDIP-seq modelling and will not describe the runtime optimized implementations of these methods.

## 3.2 Quality control metrics

### 3.2.1 Saturation analysis

The saturation analysis addresses the question, whether the number of available short reads is sufficient to generate a saturated and reproducible methylation profile of the reference genome. Figure 3.2 shows an artificial example of genomic regions with varying densities of methylated CpGs. MeDIP-seq aims to reconstruct such methylation profiles on the basis of local short read coverages. It is supposed that an insufficient number of short reads will not represent the true methylation profile. Only when a sufficient number of short reads is generated, the resulting genome vector will represent a saturated methylation profile (see Figure 3.2 A).

The basic assumption of the saturation analysis is that only a sufficient number of short reads will result in a genome wide methylation profile which will be reproducible by another independent set of a similar number of short reads. Figure 3.2 B illustrates that an insufficient number of short reads will result in methylation profiles that cannot be reproduced by another independent set of an insufficient number of short reads (Figure 3.2 B low correlation). The correlation of two independently generated genome vectors will increase when the total number of short-reads considered for the construction of each of the two genome vectors increases (Figure 3.2 B high correlation). It is supposed that the increase of the correlation between two independently generated genome vectors will

# 3 Modelling of MeDIP-seq data

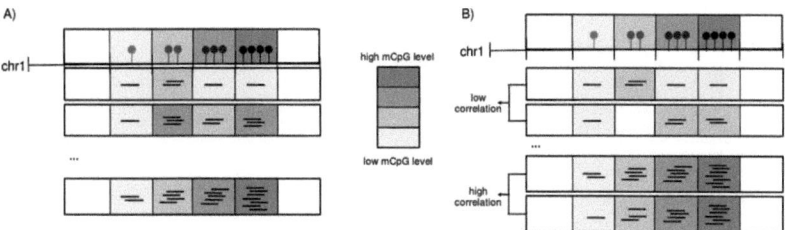

**Figure 3.2: Saturation analysis.** The figure shows an artificial example of genomic regions with varying densities of methylated CpGs. MeDIP-seq aims to reconstruct such methylation profiles on the basis of local short read coverages. Short reads are indicated as horizontal black lines. **A** It is supposed that an insufficient number of short reads will not represent the true methylation profile (see second and third row). Only when a sufficient number of short reads is generated, the resulting genome vector will represent a saturated methylation profile (last row). **B** Comparing two genome vectors generated by two independent sets of short reads will result in a low correlation when the number of short reads is not high enough to represent the methylome. It is supposed that such a correlation will increase by an increasing number of considered short reads.

saturate as soon as the total number of considered short reads is increased to a level that is able to represent the tested methylome in a saturated way. Obviously, the number of short reads that have to be generated for an sufficient sequencing depth depends on the size of the reference genome.

For the saturation analysis, the total set of available regions $(n)$ is divided into two distinct random sets $A$ and $B$ of equal size (for simplicity, both sets contain $\lfloor (\frac{n}{2}) \rfloor$ reads). Both sets $A$ and $B$ are again divided into $k$ random subsets of equal size:

$$A = \{a_1, ..., a_k\}$$

$$B = \{b_1, ..., b_k\}$$

Again for simplicity, each subset $a_1, ..., a_k$ and $b_1, ..., b_k$ contains $\lfloor \frac{(\frac{n}{2})}{k} \rfloor$ distinct randomly selected short reads. The saturation analysis runs in $k$ iterations. For each set $A$ and $B$ independently, the saturation analysis iteratively selects an increasing number of subsets and creates according genome vectors (see section 3.1) by using an arbitrary bin

**Algorithm 3.2 Simplified pseudo code for the k-iterations of the saturation analysis.**

```
for(i in 1:k){
        GA = CreateGenomeVector(a1,..., ai)
        GB = CreateGenomeVector(b1,..., bi)

        correlation=cor(GA, GB)
}
```

size (e.g. 50bp) and by previously extending the short reads to a suitable length (e.g. 400bp). In each iteration step, the resulting genome vectors $GA$ and $GB$ for the subsets of $A$ and $B$ are compared using Pearson correlation. These iterations of the saturation analysis can be noted as given in algorithm 3.2.

Correlation aims to analyse the statistical relationship between the obtained genome vectors. It is defined as

$$\text{cor}_{GA,GB} = \frac{\text{cov}(GA,GB)}{\sigma_{GA}\sigma_{GB}} \quad (3.2)$$

where $\text{cov}(GA,GB)$ is the covariance and $\sigma_{GA}$ and $\sigma_{GB}$ are the standard deviations of the temporary genome vectors $GA$ and $GB$. Let $m$ be the length of the genome vectors, and let $a_j$ and $b_j$ the j-th elements of the genome vectors $GA$ and $GB$, then the empirical correlation coefficient is defined as:

$$\text{cor}_{GA,GB} = \frac{\sum_{j=1}^{m}(a_j - \mu_{GA})(b_j - \mu_{GB})}{\sqrt{\sum_{j=1}^{m}(a_j - \mu_{GA})^2 \cdot \sum_{j=1}^{m}(b_j - \mu_{GB})^2}} \quad (3.3)$$

where

$$\mu_{GA} = \frac{1}{m}\sum_{j=1}^{m} a_j \quad (3.4)$$

$$\mu_{GB} = \frac{1}{m}\sum_{j=1}^{m} b_j \quad (3.5)$$

are the mean values of the genome vectors $GA$ and $GB$. The variables $GA$ and $GB$ are positively or negatively correlated, respectively, if $\text{cor}_{GA,GB} > 0$ or $\text{cor}_{GA,GB} < 0$ and it is valid $\text{cor}_{GA,GB} \in [-1,1]$. The closer the correlation coefficient is to either $-1$ or $1$, the stronger the correlation between the variables. If the variables are independent, this correlation coefficient is 0, but the converse is not true because the correlation coefficient detects only linear dependencies between two variables. Non-linear dependencies between two variables can be made tangible by transforming all data values into their rank representations before calculating the correlation. Such rank correlation coefficient measures the extent to which, as one variable increases, the other variable tends to increase (or decrease, respectively), without requiring that increase (or decrease, respectively) to be represented by a linear relationship.

In each of the $k$ iteration steps of the saturation analysis, two genome vectors are constructed, each one consisting of $i \cdot \lfloor \frac{\binom{n}{2}}{k} \rfloor$ randomly selected distinct short reads, where $i = 1, ..., k$. In each iteration step, the correlation of the two resulting genome vectors is calculated and stored. Instead of calculating the correlation, rank correlation can be calculated. Although it is not expected to observe non-linear dependencies, the rank correlation is more robust against outlier values, i.e. sequencing artefacts resulting in genomic bins associated to an extraordinary coverage in all experimental approaches. Such extreme outliers present in both genome vectors result in an overestimated good correlation. In addition, it might be worthwhile to calculate correlation (or rank correlation) for non-zero genomic bins, only. That means, before correlation is calculated, all values are removed from the two genome vectors where it is valid $a_j = b_j = 0$. With this, genomic bins that are not covered by any short read, e.g. because no cytosines are present in their surrounding, will not influence the resulting correlation. Both modifications for calculating the correlation can be enabled by according parameters of the presented MEDIPS package (see section 3.5).

As the number of considered short reads increases during each iteration step, it is supposed that the resulting genome vectors become more similar, a dependency that is expressed by an increased correlation. By storing the resulting correlation coefficient after each iteration step, the change of correlation during the $k$ iteration steps can be visualized by plotting the number of considered reads against the resulting correlation coefficients (see Figure 3.3). It is supposed that the correlation will approach a maximum close to 1 when a saturated number of short reads is considered. Such a plot allows for gaining an impression of the reproducibility of constructing a methylome with respect to the number of considered short reads and with respect to the size of the reference genome.

## Estimated saturation analysis

A saturation analysis can be performed on two independent sets of short reads, only. Therefore, a true data based saturation analysis can only be calculated for half of the available short reads. That is generating two genome vectors where each one contains $\lfloor(\frac{n}{2})\rfloor$ distinct short reads. Obviously, it is of interest to examine the reproducibility of the MeDIP-seq experiment for the total amount of available short reads, that is comparing two genome vectors where both are generated from $n$ short reads. But when considering the same $n$ reads for creating two genome vectors, the resulting correlation will always be 1. Therefore, a true saturation analysis can be calculated for $\lfloor(\frac{n}{2})\rfloor$ short reads, only. However, a saturation analysis can be extended by a subsequent estimated saturation analysis.

For the estimated saturation analysis, the full set of given regions is artificially doubled by considering each short read twice. Afterwards, the described saturation analysis is performed on the artificially doubled set of regions. Because the artificially doubled set of short reads does not represent a true outcome of a MeDIP-seq experiment but considers each read twice, the calculated correlations will overestimate the true reproducibility. It is assumed that the true correlation for the full set of available short reads will be between the results of the true and of the estimated saturation analysis.

## Random iterations

Methods that randomly select data entries, like for the saturation analysis, can be processed several times in order to obtain more stable results. It is recommended to perform the saturation analysis, and therefore the random partitioning of the short reads into the several subsets of $A$ and $B$, several times. The $i$-th correlation coefficient as calculated during the $k$ iteration steps of each individual saturation analysis will be finally

## 3 Modelling of MeDIP-seq data

replaced by the average of the *i-th* correlation coefficients calculated during the several repetitions of the saturation analysis. Let $cor_{(GA,GB)_{i,j}}$ be the correlation calculated in the *i-th* iteration of a saturation analysis, where $i = 1, ..., k$, and of the *j-th* random repetition of the saturation analysis, where $j = 1, ..., s$, and $s$ is the number of performed random repetitions, then the final correlation coefficient at each of the $k$ iteration steps is

$$cor_{(GA,GB)_i} = \frac{1}{s}\sum_{j=1}^{s} cor_{(GA,GB)_{i,j}} \qquad (3.6)$$

The MEDIPS package (see section 3.5) allows to specify an according parameter for executing the saturation analysis several times before the averaged results are finally stored and plotted.

## Results

Figure 3.3 A shows the results of a saturation analysis (blue curve, 10 random saturation analysis iterations) of approximately 25 million MeDIP-seq derived short reads. The maximal obtained correlation is 0.88 as derived from comparing the two genome vectors, each generated from approximately 12.5 million randomly selected distinct short reads. The estimated saturation analysis (red curve) shows the estimated maximal correlation of 0.94 as derived from the artificially doubled set of short reads. The plot illustrates similar curve shapes for the true and estimated saturation analysis. Therefore, it is expected that the correlations calculated for the true saturation analysis will show a similar behaviour to the correlations calculated for the estimated saturation analysis, in case appropriate more distinct short reads will be available. However, as mentioned above, it is expected that the estimated correlation overestimates the results for a true saturation analysis and therefore, the theoretical maximal correlation for the full set of available short reads will be between the maximal reached correlation of the saturation and of the estimated saturation analysis (that is between 0.88 and 0.94 in the given example).

Figure 3.3 B illustrates the results of the saturation analysis calculated for approximately 20 million input derived short reads (this is sequencing the sonicated DNA fragments without preceding immunoprecipitation). Here, the complexity of available DNA-

3 Modelling of MeDIP-seq data

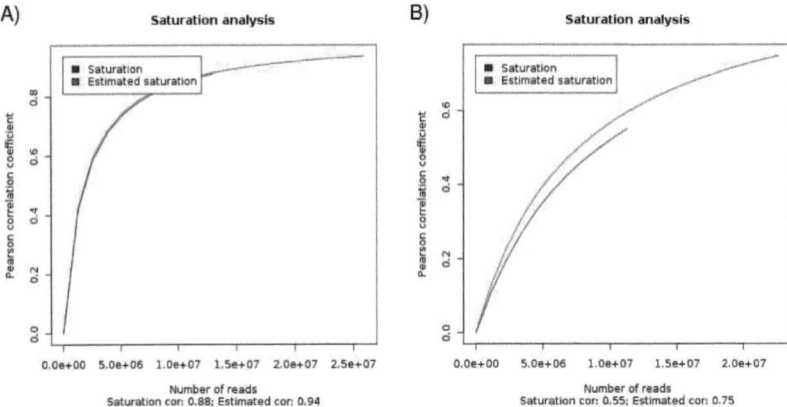

Figure 3.3: **Results of the saturation analysis.** A Results of the saturation (blue curve) and estimated saturation (red curve) analyses (performed with 10 random iterations) of approximately 25 million MeDIP-seq derived short reads. It is expected that the theoretical maximal correlation for the full set of available short reads will be between the maximal reached correlation of the saturation (0.88) and of the estimated saturation analysis (0.94). The results indicate that the human methylome of the examined biological material (here hESCs) can be well reproduced using 25 million short reads. B Results of the saturation analysis calculated for approximately 20 million input derived short reads. The maximal obtained correlation is 0.55 for the saturation and 0.75 for the estimated saturation analysis. Moreover, the estimated saturation analysis clearly overestimates the results of the saturation analysis. The results indicate that the number of input reads is far away from being able to generate reproducible full genome short read coverages.

fragments is much higher than for an immunoprecipitated sample. The figure shows that the number of available input reads is not sufficient for creating well reproducible genome vectors. The maximal reached correlation is 0.55 for the saturation and 0.75 for the estimated saturation analysis. Moreover, here the estimated saturation analysis clearly overestimates the results of the saturation analysis.

In summary, the saturation analysis reveals that an amount of 25 million MeDIP-seq derived short reads is sufficient to generate reproducible human methylomes. This number is supposed to vary for different reference genomes and DNA methylation patterns. On the other hand, 20 million input sequencing derived reads are far away from resulting in well reproducible full genome short reads coverages. The saturation analysis allows for examining the available produce of experimental specific sequencing data in order to assess, if sufficient short reads have been produced. Therefore, the saturation analysis is not restricted to be applied to MeDIP-seq data only, but other sequencing derived data, like e.g. ChIP-Seq data, can be evaluated by this approach.

### 3.2.2 Coverage analysis

The coverage analysis addresses the essential question about the genome wide depth of sequence pattern (e.g. cytosine or CpG) coverage by an increasing number of integrated sequencing derived short reads. For this, first all genomic coordinates of the sequence pattern of interest have to be identified.

Here, it is important to consider the fact that DNA sequence patterns can be reverse complementary. For example, the CpG pattern is reverse complementary and therefore exists on both strands of the DNA at the same time. On the other hand, when looking for non reverse complementary sequence patterns, both strands of the DNA have to be considered. For example, when searching for genomic positions of the single base DNA pattern C, either both strands have to be accessed or only one strand has to be accessed but here all C and G positions have to be returned. The MEDIPS package provides a function that considers this circumstance when identifying the genomic positions of arbitrary sequence patterns (see section 3.5).

**Algorithm 3.3 Simplified pseudo code for the k-iterations of the coverage analysis.**

```
for(i in 1:k){
        coverages=CalculateCoverage(A=(a1,..., ai), P=(p1,...,pm)
        , Q=(q1,..., ql))
}
```

In the following, it is expected that all genomic pattern positions are stored on a vector $P = (p_1,...,p_m)$ where $m$ is the number of sequence patterns present in the reference genome. For the coverage analysis, the total set of available short reads ($A$) is divided into $k$ random subsets of equal size:

$$A = \{a_1,...,a_k\}$$

For simplicity, each subset $a_1,...,a_k$ contains $\lfloor (\frac{n}{k}) \rfloor$ distinct randomly selected short reads, where $n$ is the total number of available short reads. The coverage analysis iteratively selects an increasing number of subsets and tests how many pattern positions from $P$ are covered by the available regions. In addition, the coverage analysis counts how many $p_i$'s are covered at least $Q$ times, where $Q = (q_1,...,q_l)$ represents an arbitrary number of coverage depths to be tested. For example, the according function of the MEDIPS package (see section 3.5) tests by default how many CpGs are covered at least 1x, 2x, 3x, 4x, 5x, and 10x (this is equivalent to the notation $Q = 1,2,3,4,5,10$). In principle, the coverage analysis can be noted as shown in algorithm 3.3.

In each iteration step, the coverage analysis selects $i$ subsets of $A$ and calculates how many $p_i$'s are covered at least $Q = (q_1,...,q_l)$ times. For each level of $Q$, the returned *coverages* object stores how many $p_i$'s are covered by the tested depths and the results of each iteration step are stored. The $k$-th iteration step of the coverage analysis shows the depth of sequence pattern coverages obtained with the full set of available short reads. The advantage of the iterative approach is that the behaviour of pattern coverage can be examined with respect to an increasing number of considered short reads. For this, coverage curves can be generated by plotting the number of covered sequence patterns for each level of $Q$ against the number of considered short reads. The progression of the resulting coverage curves indicate the state of saturation of the overall sequence pattern

## 3 Modelling of MeDIP-seq data

coverages. As for calculating the genome vector (see section 3.1) or as for the saturation analysis (see subsection 3.2.1), it is recommended to previously extend the short read lengths to e.g. 400bp.

## Random iterations

Methods that randomly select data entries, like for the coverage analysis, can be processed several times in order to obtain more stable results. It is recommended to perform the coverage analysis, and therefore the random partitioning of the short reads into the several subsets of $A$, several times. The $i$-th observed coverages for the tested levels of $Q$ as calculated during the $k$ iteration steps of each individual coverage analysis will be finally replaced by the average of the $i$-th correlation coefficients as calculated during the several random repetitions of the saturation analysis. Let $coverages_{i,j}$ be the coverage of any specified level of $Q$ as calculated in the $i$-th iteration of a coverage analysis, where $i = 1,...,k$, and of the $j$-th random repetition of the coverage analysis, where $j = 1,...,s$, and $s$ is the number of performed random repetitions, then the final coverage at each of the $k$ iteration steps is

$$coverages_i = \frac{1}{s}\sum_{j=1}^{s} coverages_{i,j} \qquad (3.7)$$

The MEDIPS package (see section 3.5) allows to specify an according parameter for executing the coverage analysis several times before the averaged results are finally stored and plotted.

## Results

As an example, Figure 3.4 shows the results of a coverage analysis for approximately 26 million MeDIP-seq derived short reads from a human sample, where the coverage of the CpG sequence pattern was examined for the levels $Q = 1, 2, 3, 4, 5, 10$ (10 random iterations). In total, the human genome (hg19 [86]) contains approx. 28.2 million CpGs. From these, approximately 22.4 million CpGs (79.3%) are covered at least one time ($q_i = 1$) when the full set of approximately 26 million unique short reads is considered. The progression of the coverage curve for the level $q_i = 1$ (red curve) indicates that the CpG coverage is

## 3 Modelling of MeDIP-seq data

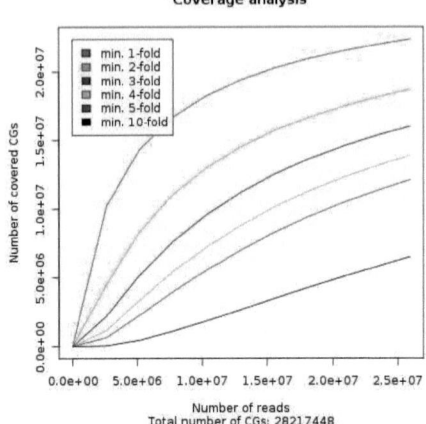

**Figure 3.4: Coverage analysis.** The Figure shows the results of a coverage analysis for approximately 26 million MeDIP-seq derived short reads from a human sample, where the coverage of the CpG sequence pattern was examined for the levels $Q = 1, 2, 3, 4, 5, 10$. In total, the human genome (hg19 [86]) contains approx. 28.2 million CpGs. From these, approx. 22.4 million CpGs (79.3%) are covered at least one time ($q_i = 1$, red curve) when the full set of approx. 26 million short reads is considered.

approaching a maximum and therefore, by adding more short reads, the coverage does not increase in a linear way anymore.

Although an increase of short reads will always improve the sequence pattern coverage, the coverage analysis, together with the saturation analysis (see subsection 3.2.1), allow for gaining an impression on the overall sequence pattern coverage and reproducibility of constructing a methylome based on the total number of MeDIP-seq derived short reads. These data quality controls assist in deciding whether the costs of additional experimental runs are in due proportion to the expected improvements of coverage and reproducibility.

### 3.2.3 CpG enrichment

As a third MeDIP-seq data quality control, the CpG enrichment approach examines how strong the genomic regions underlying the obtained short reads are enriched for CpGs compared to the frequency of CpGs within the reference genome.

# 3 Modelling of MeDIP-seq data

For this, first the number of Cs ($G.c$), the number of Gs ($G.g$), the number CpGs ($G.cg$), and the total number of bases ($m$) within the specified reference genome are counted. Subsequently, the relative frequency of CpGs and the observed/expected ratio of CpGs [32] as present in the reference genome are calculated as:

$$Genome.CpG_{rel.f} = \frac{G.cg}{m} \qquad (3.8)$$

$$Genome.CpG_{obs/exp} = \frac{G.cg \cdot m}{G.c \cdot G.g} \qquad (3.9)$$

Additionally, the number of Cs ($SR.c$), the number of Gs ($SR.g$), the number CpGs ($SR.cg$), and the total number of bases ($n$) are counted for the DNA sequences underlying the given short reads, only. Subsequently, the relative frequency of CpGs and the observed/expected ratio of CpGs as present in the short reads specific DNA sequences is calculated accordingly:

$$SR.CpG_{rel.f} = \frac{SR.cg}{n} \qquad (3.10)$$

$$SR.CpG_{obs/exp} = \frac{SR.cg \cdot n}{SR.c \cdot SR.g} \qquad (3.11)$$

The final enrichment values result by dividing the relative frequency of CpGs (or the observed/expected value, respectively) of the regions by the relative frequency of CpGs (or the observed/expected value, respectively) of the reference genome:

$$enrich_{rel.f} = \frac{SR.CpG_{rel.f}}{Genome.CpG_{rel.f}} \qquad (3.12)$$

| - | #reads (m) | # C (m) | #G (m) | # CpGs (m) | Rel. freq. | obs/exp | E. rel. freq. | E. obs/exp |
|---|---|---|---|---|---|---|---|---|
| Genomic hg19 | - | 592.4 | 592.8 | 28.6 | 0.9904 | 0.2363 | - | - |
| MeDIP, hESC | 25.9 | 231.3 | 232.2 | 20.0 | 2.0884 | 0.3575 | **2.1087** | 1.5129 |
| MeDIP, DE | 32.5 | 298.1 | 299.4 | 30.8 | 2.5618 | 0.4168 | **2.5867** | 1.7638 |
| INPUT, hESC | 6.2 | 54.1 | 54.3 | 3.0 | 1.3276 | 0.2432 | **1.3405** | 1.0290 |
| INPUT, DE | 16.3 | 132.3 | 132.9 | 6.5 | 1.0771 | 0.2234 | **1.0876** | 0.9453 |

**Table 3.2: CpG Enrichment.** The table shows the results of the CpG enrichment analysis of the MEDIPS package. CpG enrichment is calculated for MeDIP-seq data from hESCs, DE and input samples (here, input samples from hESCs and DE were processed separately) relative to the reference genome (hg19). m=million, E. rel. freq=CpG enrichment relative frequency, E. obs/exp= CpG enrichment observed/expected.

$$enrich_{obs/exp} = \frac{SR.CpG_{obs/exp}}{Genome.CpG_{obs/exp}} \quad (3.13)$$

For short reads derived from an input experiment (that is sequencing of none-enriched DNA fragments), the enrichment values are expected to be close to 1. In contrast, short reads derived from MeDIP-seq experiments are expected to be enriched for CpG rich DNA sequences that will be indicated by increased enrichment scores. Table 3.2 shows the results of CpG enrichments calculated for two MeDIP and for two input sequencing derived data sets. Enrichment scores show that MeDIP derived short reads are enriched for CpG's (relative frequency is 2.10 and 2.59, respectively), whereas input derived short reads do not show comparable CpG enrichments (relative frequency is 1.34 and 1.09, respectively).

## 3.3 Normalization

The idea of a MeDIP experiment is to identify cytosine methylation profiles of a sample of interest by immunocapturing methylated cytosines using a mCpG specific antibody [103]. However, it has been shown [27, 80] that MeDIP signals scale with local densities of CpGs and are not necessarily influenced by methylated cytosines, only. Therefore, the need for MeDIP-seq data correction occurs through an unspecific binding of the utilized antibody to unmethylated cytosines, especially in the context of low densities of methylated cytosines.

## 3.3.1 Reads per million (rpm)

For each pre-defined genomic bin, the genome vector (see section 3.1) stores the number of provided overlapping extended short reads and these are interpreted as the raw MeDIP-seq signals. Based on the total number of provided short reads ($n$), the raw MeDIP-seq signals can be transformed into a reads per million ($rpm$) format in order to assure that coverage profiles derived from different biological samples are comparable, although generated from differing amounts of short reads. Let $x_{bin_i}$ be the raw MeDIP-seq signal of the genomic bin $i$, where $i = 1,...,m$ and $m$ is the total number of genomic bins, then the $rpm$ value of the genomic bin is simply defined as:

$$rpm_{bin_i} = \frac{x_{bin_i} \cdot 10^6}{n} \qquad (3.14)$$

The MEDIPS package (see section 3.5) allows for exporting wiggle files [86] containing genome wide $rpm$ values at an user-specified resolution (e.g. 50 bp). By utilizing these wiggle files, the $rpm$ profiles of the processed biological sample can be immediately visualised using a suitable genome browser.

## 3.3.2 Coupling factors

Similar to other MeDIP normalization approaches [27, 80], the presented method corrects for the unspecific antibody binding by incorporating local CpG densities into the MeDIP-seq derived signals. In order to integrate the information about CpG densities into the following analysis, it is necessary to identify the genomic positions of all CpGs. This can be achieved by executing the *MEDIPS.getPosition()* function of the MEDIPS package (see section 3.5). Following the valuable concept of coupling factors presented by Down et al. [27], a *coupling vector* is calculated based on the received genomic positions of all CpGs.

The coupling vector is of the same size as the predefined genome vector (see section 3.1) but contains local CpG densities (also called coupling factors) for each genomic bin, instead. For each predefined genomic bin at position $b$, the density of surrounding CpGs

## 3 Modelling of MeDIP-seq data

has to be calculated. For this, first a maximal distance ($d$) has to be defined. Only CpGs within the range of $[b-d, b+d]$ will contribute to the final local coupling factor at $b$. The optimized value for $d$ will reflect the estimated size of the sonicated DNA fragments after amplification and size selection. This is because MeDIP-seq derived signals at position $b$ are influenced by sequenced DNA fragments that overlap with position $b$. Immunoprecipitation of these DNA fragments can be caused by a methylated and antibody bound CpG located at any position of the DNA-fragment. The maximal distance of a CpG contributing to the signal at $b$ is therefore the estimated length of the sonicated DNA fragments ($d$).

There are several ways of calculating coupling factors for genomic bins. Let $c$ be the chromosomal position of a CpG and as $b$ is the chromosomal position of a genomic bin, $dist = |b-c|$ is the distance between the genomic bin and the CpG. A CpG will contribute to the coupling factor of a genomic bin at position $b$, if $dist \leq d$. The simplest way is to count the number of CpGs within the maximal defined distance $d$ around a genomic bin at position $b$ (*count* function). Another approach is to weight each CpG by its distance to the current genomic bin. CpGs farther away from the current genomic bin will receive smaller weights, whereas CpGs close to the genomic bin will receive higher weights.

The upper panel in Figure 3.5 illustrates a genome vector generated by defining a bin size of 50bp. In addition, CpGs are given in a schematic way. The figure shows that immunoprecipitated DNA fragments of an estimated average length greater than the defined bin size can contribute to the signal of a genomic bin at position $b$. Moreover, the schematic distance function illustrates that CpGs close to position $b$ will receive higher weights than CpGs located farther away.

There are several possible ways for defining weighting functions. In the context of this work, the following weighting functions were evaluated: *count*, *linear*, *exp* [80], *log* [80], and *custom* [27]. Algorithm 3.4 shows appropriate source code for implementation of these weighting functions. In addition, the images at the bottom of Figure 3.5 show the progression of these weighting functions by defining a maximal distance $d = 700$.

Whereas the weighting functions *count*, *linear*, *exp*, and *log* are calculated by the given

## 3 Modelling of MeDIP-seq data

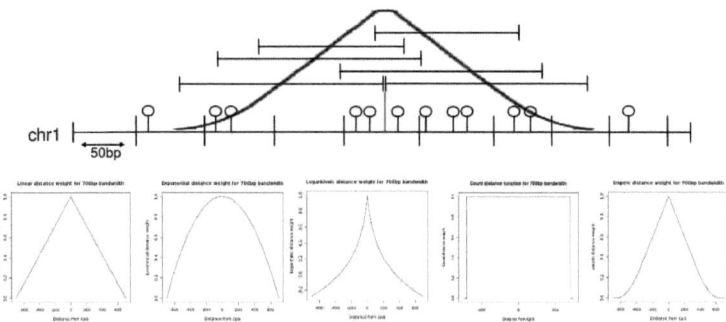

**Figure 3.5: Calculation of coupling factors.** The upper panel shows a schematic view of the genome vector created by defining a bin size of 50bp. In addition, CpGs are shown in a schematic way. A coupling factor is calculated for the centered genomic bin at position $b$ (marked by a red vertical line). For this, all CpGs within a maximal distance $d$ are considered. The maximal distance $d$ reflects the estimated average size of sequenced DNA fragments. There are several ways for calculating coupling factors. The simplest way is to count the number of CpGs in the surrounding of $b$ but with a maximal distance of $d$. Alternatively, a weighting function can be applied in order to weight each CpG by its distance ($dist$) to the current genomic bin at position $b$. There are several possible weighting function. The five images at the bottom of the Figure show the progression of the weighting functions *linear*, *exp*, *log*, *count*, and *custom* [27] by defining $d = 700$.

---

**Algorithm 3.4 Distance functions for calculating coupling factors.** Supported distance functions are *count*, *linear*, *exp* [80], *log* [80], and *custom* [27]. Here, $d$ is the maximal distance around a genomic bin where CpGs are considered at all, and $dist$ is the actual absolute distance between a genomic bin and a CpG.

```
if(func=="count"){wfun=function(dist,d){if(d<=dist){return(1)}
   else{return(0)}}}

if(func=="linear"){wfun=function(dist,d){if(d<=dist){return(1-
   dist/d)}else{return(0)}}}

if(func=='exp'){wfun=function(dist,d){if(d<=dist){return(1-dist
   ^2/(d)^2)}else{return(0)}}}

if(func=='log'){wfun=function(dist,d){if(d<=dist){return(1-log
   (1+abs(dist)/(d/18),10))}else{return(0)}}}

if(func=='custom'){wfun=function(dist){if(dist<=(length(
   dFileData[,1])-1)){return(dFileData[(dist+1),2])}else{return
   (0)}}}
```

formulas, the *custom* function allows for specifying user-defined weights for any possible distance *dist*. For example, Down et al. [27] have generated custom weights for the distances $dist \in [0, 648]$. These weights were estimated empirically by sampling from the fragment-length distribution and randomly placing each fragment such that it overlaps the genomic bin [27]. These weights are stored in an array (here *dFileData*) and are returned when the *custom* function is called with a given pre-calculated *dist* value.

Let $C_{cb}$ be the coupling factor between a CpG at position $c$ and a genomic bin at position $b$ calculated based on an arbitrary weighting function and for any specified parameter $d$. Then

$$C_{tot} = \sum_c C_{cb} \qquad (3.15)$$

is the sum of coupling factors at the genomic bin $b$ with respect to all CpGs at a genomic position $c$, where $|b-c| \leq d$. For simplification, in the following, $C_{tot}$ is called the coupling factor at a genomic bin $b$ and gives a (weighted) measure of local CpG density.

It has been shown [103, 30] that in mammalian cells, methylation is negatively correlated to CpG densities. In other words, regions of low CpG density tend to be highly methylated, whereas regions of high CpG density tend to be mainly unmethylated. In order to test the correlation of measured methylation values [30] compared to local CpG densities calculated with respect to the different weighting functions, I have systematically calculated coupling vectors (bin size=50) with varying $d \in [0, 2000]$ using the weighting functions *count, linear, exp, log,* as well as for the empirically derived weights presented by Down et al. [27] (*custom*). Because the *custom* weights are available for the range $dist \in [0, 648]$, only, the weight at $dist = 648$ is also utilized for the remaining distances up to $d = 2000$.

For the comparisons, I have accessed DNA-methylation values derived from bisulphite sequencing experiments of a sperm sample as presented by the human epigenome project (HEP) [30]. Bisulphite sequencing derived methylation data was generated for approximately 3000 selected genomic regions (called HEP traces) of length 50bp to 500bp [30]. In order to compare CpG densities to the available methylation data, for all utilized weighting functions with varying parameter $d$, we have calculated mean coupling factors for each

## 3 Modelling of MeDIP-seq data

of the HEP traces and examined the relation to corresponding mean methylation values by Pearson correlation.

Figure 3.6 a shows the resulting Pearson correlation for varying parameter $d$ and for the several tested weighting functions. Interestingly, the best negative correlation (-0.73; that is the higher the CpG density, the lower the bisulphite derived methylation values) was achieved by setting the parameter $d = 700$ and by using the *count* function. For this parameter settings, Figure 3.6 b shows a scatter plot comparing mean HEP methylation values and mean coupling factors. Here, each data point represents a HEP trace and the plot contrasts the mean methylation value (x-axis) to the mean CpG density (y-axis). The colour code divides the full range of CpG densities into four regular quantiles. Based on these results, in the following, the coupling vector is always calculated by specifying $d = 700$ and by using the *count* function. However, the MEDIPS package (see section 3.5) allows for justifying the according parameters or for supplying any custom defined distance weights.

Coupling vectors can be calculated for any arbitrary DNA sequence pattern using the MEDIPS package (see section 3.5). Moreover, the resulting coupling vectors can be exported into a wiggle file [86] that allows for visualizing the sequence pattern densities along the chromosomes using a suitable genome browser.

### 3.3.3 Calibration curve

As I have created a genome vector that contains the raw signals at each genomic bin as well as an according coupling vector containing the calculated coupling factors at each genomic bin, the dependency of local MeDIP-seq signal intensities and local CpG densities can be examined. However, by simply plotting the genome vector against the coupling vector, no concrete dependency is observable (see Figure 3.7). Nevertheless, it can already be observed that increased MeDIP-seq signals are preferential present in the low range of CpG coupling factors. Genomic bins associated to higher CpG densities show comparatively lower MeDIP-seq signals.

However, a dependency between CpG densities and MeDIP-seq signals can be made

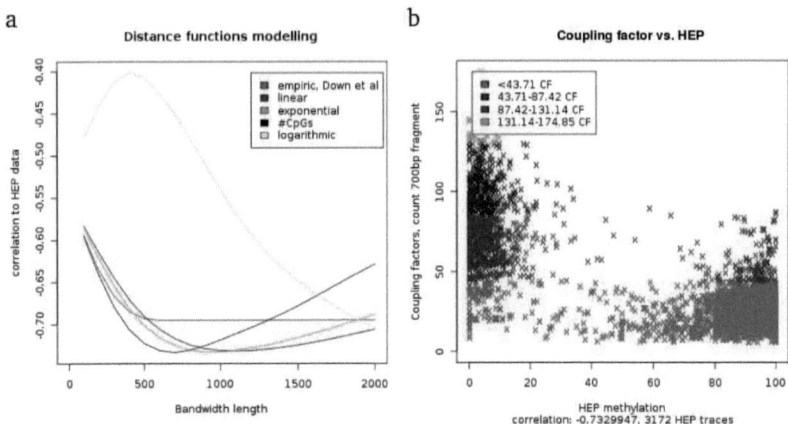

**Figure 3.6: Evaluation of coupling factor calculations.** Panel **a** shows the resulting Pearson correlations (y-axis) between the mean coupling factors and bisulphite sequencing derived mean methylation values for a varying distance parameter $d$ (x-axis) and for different weighting factors (colours). The best negative correlation (-0.73) was achieved by setting the parameter $d = 700$ and by using the *count* function. Panel **b** shows the according scatter plot where each data point represents a HEP trace. The scatter plot contrasts the mean methylation value (x-axis) with the mean CpG density (y-axis). The color code divides the full range of CpG densities (coupling factors, CF) into four regular quantiles.

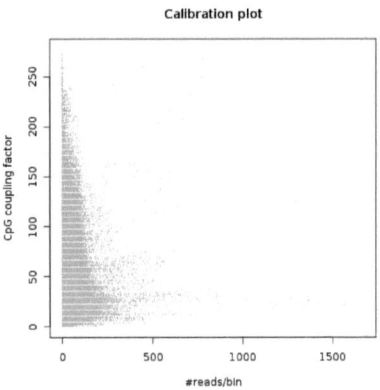

**Figure 3.7: Full range calibration plot.** The raw MeDIP-seq derived signals (genome vector, x-axis) is compared to the CpG coupling factors (coupling vector, y-axis). Each data point represent a genomic 50bp bin. When looking on the full range of MeDIP-seq signals, no general dependency is observable.

## 3 Modelling of MeDIP-seq data

**Algorithm 3.5 Calculation of the calibration curve.** Column three of the *genomeVecObj* contains the pre-calculated MeDIP-seq signals and the third column of the *couplingVecObj* contains the pre-calculated coupling factors of the genomic bins.

```
signal=as.numeric(genomeVecObj$genomeVec[,3])
coupling=as.numeric(couplingVecObj$genomeCoup[,3])
maxCoup=floor(max(coupling))

mean_signal=rep(NA, times = maxCoup+1)
mean_coupling=rep(NA, times = maxCoup+1)

for(i in 0:maxCoup){
        mean_signal[i+1]=mean(signal[coupling>=i & coupling<(i
            +1)])
        mean_coupling[i+1]=mean(coupling[coupling>=i & coupling
            <(i+1)])
}
```

tangible by calculating the calibration curve. Calculation of the calibration curve is achieved by first dividing the total range of coupling factors into regular levels. Second, all genomic bins are partitioned into these levels by considering their associated coupling factors. Finally, for each level of coupling factors, the mean signal and mean coupling factor of all genomic bins that fall into this level are calculated.

Algorithm 3.5 shows according R code for calculating the calibration curve. For this, the maximum of all pre-calculated coupling factors (see subsection 3.3.2) is extracted (here *maxCoup*) and is utilized for separating the full range of coupling factors into regular levels ranging from 0 to *maxCoup*. The loop runs over the full range of coupling factors and iteratively selects the genomic bins associated to the current level of coupling factors. In each iteration step, the mean signal and mean coupling factors of all temporarily selected genomic bins are calculated. Finally, for each processed level of coupling factors, the calibration curve contains the mean coupling factors and mean MeDIP-seq signals.

As the calibration curve represents the averaged signals and coupling factors over the full range of coupling factors, it reveals the experiment specific dependency between signal intensities and CpG densities. Figure 3.8 a shows the scatter plot as shown in Figure 3.7 but with an X-axis range limit of 50 reads/bin. Therefore, the plot only shows genomic bins associated to MeDIP-seq signals of max. 50 reads per bin but it contains the full

# 3 Modelling of MeDIP-seq data

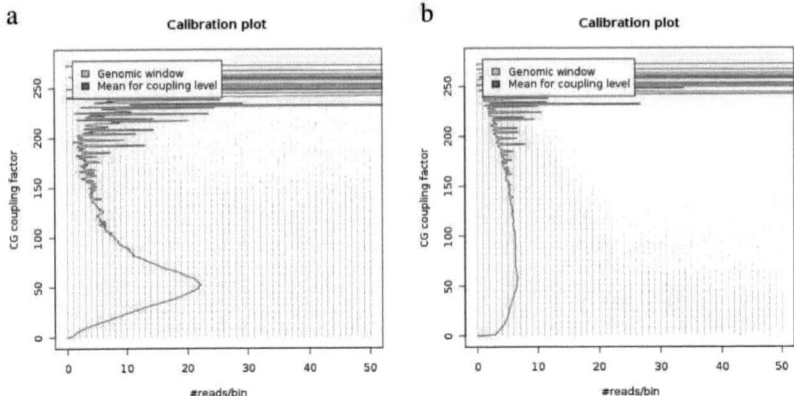

**Figure 3.8: Calibration curves for MeDIP-seq and INPUT-Seq data.** Each data value within a calibration plot represents a genomic bin. The x-axis shows the raw signals and the y-axis shows the coupling factors for the genomic bins. The plots include genomic bins associated to MeDIP-seq signals $\leq 50$ overlapping short reads per bin, only. The calibration curve (red curve) is very characteristic for MeDIP-seq experiments (see panel **a**). For low levels of CpG coupling factors, the calibration curve illustrates that the MeDIP-seq signals, in average, increase just because of an increasing CpG density. An analogous dependency is, as expected, not observable for INPUT derived sequencing data (see panel **b**). The noise of the calibration curve in the high range of coupling factors results by a decrease of the number of genomic bins associated to high levels of coupling factors. Therefore, the calculated means for high levels of coupling factors are influenced by single outliers.

range of coupling factors. Moreover, the plot contains the pre-calculated calibration curve (red line).

In fact, for the low range of coupling factors, the calibration curve in Figure 3.8 a indicates that the MeDIP-seq signals, in average, increase because of an increasing CpG density. Therefore, an increased signal is not necessarily caused by a higher level of methylated cytosines but scales with the general CpG density. In contrast, Figure 3.8 b shows the calibration curve calculated for INPUT derived sequencing data. Here, the described dependency of CpG density and sequencing signals is not observable. Therefore, the calibration plot is very characteristic for MeDIP data and the quality of the enrichment step of the MeDIP experiment can be estimated by visual inspection of the progression of the calibration curve.

For higher levels of CpG densities, the mean MeDIP-seq signals decrease (see Figure 3.8 a). It is assumed that this decrease is caused by the fact that in biological systems, regions of higher CpG densities are mainly unmethylated. Interestingly, in biological systems, cytosine methylation occurs mainly in regions of low CpG density. In contrast, cytosines located in regions of high CpG density are mainly unmethylated. This circumstance implicates that the dependency between increased signal intensities caused by increased CpG densities is visible for regions of low CpG densities, only.

### 3.3.4 Relative methylation score (rms)

The calibration curve (see subsection 3.3.3) reveals that, on average, an increase of MeDIP-seq signals is caused by an increasing CpG density. This approximately linear dependency is visible for the low range of coupling factors, only (see Figure 3.8 a). For higher levels of CpG densities, the mean MeDIP-seq signals decrease. As mentioned above, it is assumed that this decrease is caused by the fact that in mammalian cells, regions of higher CpG densities are mainly unmethylated.

In agreement with this assumption, Pelizzola and colleagues [80] have shown that the dependency of MeDIP derived signals and CpG density continues for higher levels of CpG densities, by analyzing artificially fully methylated samples using MeDIP-Chip. Figure 3.9 illustrates their identified sigmoidal dependency between CpG density (x-axis) and MeDIP-Chip data (y-axis). In agreement with Pelizzola et al., it is assumed that the signal plateau in the lower range of chip signals is caused by background noise but in contrast to Pelizzola et al., it is assumed that the signal plateau in the upper range of chip signals occurs by a saturation of hybridization events and is therefore an array specific artefact.

Motivated by the discussed observations made by Pelizzola et al. [80] and by visual inspection of the MeDIP-seq derived calibration curve (see Figure 3.8 a), a continuing linear dependency of MeDIP-seq signals for higher levels of CpG densities is assumed.

Analogous to Down et al. [27], the local maximum of mean MeDIP-seq signals of the

3 Modelling of MeDIP-seq data

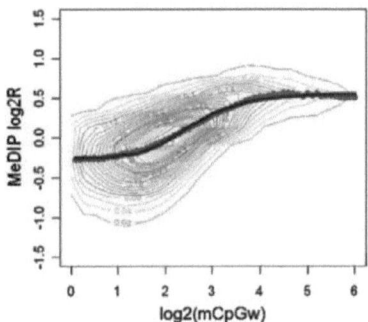

**Figure 3.9: MeDIP-Chip signals as a logistic function of the methylation level [80].** Pelizzola and colleagues observed a sigmoidal dependency between CpG density (x-axis, mCpGw) and MeDIP-chip signals (y-axis, *MeDIP log2R*) using artificially fully methylated samples. Their signal ratios *(R)* are calculated by dividing each tiling array derived probe signal of the MeDIP sample by the corresponding probe signal of an Input sample. For all probes on the tiling array, Pelizzola et al. have calculated weighted CpG densities (*CpGw*). Because here they have analyzed artificially fully methylated samples by the MeDIP approach, it is assumed that all CpGs are methylated, i.e. *CpGw=mCpGw* [80]. From this it follows that all CpGs in the surrounding of a probe will contribute to the MeDIP-chip signal. Therefore, the presented sigmoidal curve represents the dependency of MeDIP-chip signals on the local weighted CpG densities over the full range of CpG densities.

# 3 Modelling of MeDIP-seq data

calibration curve in the lower part of coupling factors is identified. Let

$$y = y_1, ..., y_l \tag{3.16}$$

be the mean coupling factors, and let

$$x = x_1, ..., x_l \tag{3.17}$$

be the according mean MeDIP-seq signals of the calibration curve, where $l$ is the number of tested coupling factor levels and $i = 1, ..., l$, then the smallest level $i$ is identified, where for all $j = i-3, i-2, i-1, i+1, i+2, i+3$ it is valid

$$x_i \geq x_j \tag{3.18}$$

Let $i_{max}$ be the according identified level of $i$, then

$$y_{max} = y_1, ..., y_{i_{max}} \tag{3.19}$$

$$x_{max} = x_1, ..., x_{i_{max}} \tag{3.20}$$

are the parts of the calibration curve in the low range of coupling factors, where an approximately linear dependency between MeDIP-seq signals and coupling factors is observed.

Here, $x_{max}$ can be explained by a function of $y_{max}$ as

$$x_{max} = f(y_{max}) + \varepsilon \tag{3.21}$$

where $\varepsilon$ is a non-deterministic error variable (i.e. measurement errors) that is expected to follow a random distribution with expectation value $E(\varepsilon) = 0$. Because linear dependency between $x_{max}$ and $y_{max}$ is assumed, $x_{max}$ can be described as

$$x_{max} = \alpha + \beta \cdot y_{max} + \varepsilon \tag{3.22}$$

where the parameter $\alpha$ is the intercept, and the parameter $\beta$ is the slope of the linear approximation. Based on the pre-calculated $x_{max}$ and $y_{max}$ vectors, linear regression

## 3 Modelling of MeDIP-seq data

is performed, in order to identify a suitable linear model. Linear regression estimates regression coefficients $a$ and $b$ for the parameters $\alpha$ and $\beta$ so that it is valid:

$$x_{max_i} = a + b \cdot y_{max_i} + e_i \tag{3.23}$$

where $i = 1, ..., i_{max}$. Here, the residuum $e_i$ reflects the difference between the regression curve $a + b \cdot y_{max_i}$ and the measurements of $x_{max_i}$. Moreover, $x_{max_i}$ can be replaced by an estimate $\hat{x}_{max_i}$, where

$$x_{max_i} - \hat{x}_{max_i} = e_i \tag{3.24}$$

and therefore, it is valid:

$$\hat{x}_{max_i} = a + b \cdot y_{max_i} \tag{3.25}$$

For estimating approximate values for the unknown parameters $\alpha$ and $\beta$, the least squares approach is utilized. In principle, the least squares approach identifies concrete regression coefficients $a$ and $b$ so that the sum of squared residues $e_i$, and therefore, the differences between the $x_{max_i}$ variables of the calculated calibration curve and the $\hat{x}_{max_i}$ variables of the resulting regression curve, are minimized as:

$$\sum_{i=1}^{i_{max}} e_i^2 = \sum_{i=1}^{i_{max}} (x_{max_i} - (a + b \cdot y_{max_i}))^2 \to min \tag{3.26}$$

In the context of linear regression, estimates for the regression coefficients $a$ and $b$ can be directly calculated by:

$$b = \frac{\sum_{i=1}^{i_{max}} (y_{max_i} - \bar{y}_{max})(x_{max_i} - \bar{x}_{max})}{\sum_{i=1}^{i_{max}} (y_{max_i} - \bar{y}_{max})^2} = \frac{S_{x_{max} y_{max}}}{S_{y_{max} y_{max}}} \tag{3.27}$$

and

$$a = \bar{x}_{max} - b \cdot \bar{y}_{max} \tag{3.28}$$

where

$$\bar{x}_{max} = \frac{1}{i_{max}} \sum_{i=1}^{i_{max}} x_{max_i} \tag{3.29}$$

$$\bar{y}_{max} = \frac{1}{i_{max}} \sum_{i=1}^{i_{max}} y_{max_i} \tag{3.30}$$

## 3 Modelling of MeDIP-seq data

are the according averages. $S_{x_{max}y_{max}}$ is the empirical covariance between $x_{max}$ and $y_{max}$, and is divided by $S_{y_{max}y_{max}}$, i.e. the empirical variance of $y_{max}$. The estimate $\frac{S_{x_{max}y_{max}}}{S_{y_{max}y_{max}}}$ is called the ordinary least square estimate (OLS). The OLS estimate given in formula 3.27 can be derived from formula 3.26 (see Appendix 1).

When having calculated estimates for the regression coefficients according to the derived OLS formula (3.27) and according to formula 3.28, concrete values for the parameter $\hat{x}_{max_i}$ can be calculated as:

$$\hat{x}_{max_i} = a + b \cdot y_{max_i} \qquad (3.31)$$

where $i = 1, ..., i_{max}$. For the low range of coupling factors, these estimates model the observed progression of the calibration curve. As discussed above, a continuing linear dependency between MeDIP-seq signals and CpG density is expected for the higher range of coupling factors. Based on the obtained linear model parameters, concrete $\hat{x}_{max_i}$ values can be calculated for the full range of coupling factors. Therefore,

$$\hat{x} = \hat{x}_1, ..., \hat{x}_{i_{max}}, ..., \hat{x}_l \qquad (3.32)$$

are the estimated mean MeDIP-seq signals over the full range of coupling factor levels $l$, calculated with respect to the obtained linear model parameters.

In the following, the obtained $\hat{x}_i$ values are considered as the expected MeDIP-seq signals of genomic bins associated to the according coupling factor levels of $i$, where $i = 1, ..., l$ and $l$ is the number of coupling factor levels. For MeDIP-seq data normalization, $\hat{x}$ is utilized in order to weight the observed MeDIP-seq signals of the genomic bins by the expected MeDIP-seq signals. This $\frac{observed}{expected}$ ratio serves as a divergence indicator that allows for estimating the strength of MeDIP-seq signal enrichments with respect to local CpG densities.

Let $(x_{bin_i}, y_{bin_i})$ be the raw MeDIP-seq signal of the genomic bin $i$ (i.e. the number of overlapping extended short reads), and the pre-calculated coupling factor at the genomic bin $i$, where $i = 1, ..., m$ and $m$ is the total number of genomic bins, then the normalized

## 3 Modelling of MeDIP-seq data

relative methylation score is defined as

$$rms_{bin_i} = \log_2(\frac{x_{bin_i} \cdot 10^6}{(a+b \cdot y_{bin_i}) \cdot n}) = \log_2(\frac{x_{bin_i} \cdot 10^6}{\hat{x}_{bin_i} \cdot n}) \qquad (3.33)$$

where $\hat{x}_{bin_i} = a + b \cdot y_{bin_i}$ is the estimated weighting parameter obtained by considering the coupling factor $y_{bin_i}$ of the genomic bin $i$, and $n$ is the total number of short reads considered for the generation of the genome vector. Based on the total number of short reads ($n$), the raw MeDIP-seq signals are, in parallel, transformed into a reads per million ($rpm$) format in order to assure that $rms$ values are comparable between methylomes generated from differing amounts of short reads.

### 3.3.5 Absolute methylation score (ams)

We consider the $rms$ values as the normalized MeDIP-seq signals corrected for the experiment specific effect of unspecific antibody binding. In order to identify an absolute methylation estimate for any specified region of interest, i.e. either any functional genomic region like promoters or CpG islands or genome wide windows of arbitrary length, the raw MeDIP-seq values can be normalized into absolute methylation scores ($ams$). The absolute methylation scores additionally correct for the relative CpG density of the region of interest and therefore, allow for comparing methylation profiles of genomic regions having different CpG densities.

This is especially necessary, when local methylation levels are associated to further functional and regulatory mechanism like e.g. gene expression alterations. As an example, it is supposed [85, 80] that methylation levels of proximal promoters influence the transcription rate of the according genes. However, promoters are known to show a wide spread spectrum of CpG densities. Therefore, a fully methylated high CpG density promoter will show much higher MeDIP signals than a fully methylated low CpG density promoter, although in both cases the promoter methylation level influences the transcription rate in a comparable way. Therefore, it remains inaccurate to conclude an absolute measure of promoter methylation by comparing MeDIP-seq derived $rpm$ or $rms$ signals from promoters having different CpG densities.

## 3 Modelling of MeDIP-seq data

Let

$$ROI = ((x_{bin_1}, y_{bin_1}), ..., (x_{bin_s}, y_{bin_s}))  \qquad (3.34)$$

be the raw MeDIP-seq signals and coupling factors of adjacent genomic bins $i$ that define a region of interest (ROI), where $i = 1, ..., s$ and $s$ is the total number of genomic bins comprised by the ROI, then the absolute methylation score for the ROI is defined as:

$$ams_{ROI} = \log_2\left(\frac{\frac{1}{s}\sum_{i=1}^{s} \frac{(x_{bin_i} \cdot 10^6)}{(a + b \cdot y_{bin_i}) \cdot n}}{\frac{1}{s}\sum_{i=1}^{s} y_{bin_i}}\right) \qquad (3.35)$$

Analogous to Pelizzola et al. [80], we interpret the $ams$ values (Pelizzola et al. call them $rms$), as the measure of the normalized methylation that is independent of the CpG density of the corresponding region.

### 3.3.6 Evaluation of MeDIP-seq data normalization

As described in section 3.3.4, the rational behind the presented normalization method is based upon the concept of coupling factors presented by Down et al. [27]. Based on a selected distance function for calculating coupling factors (see section 3.3.2), we estimated the dependency between increasing total CpG density and increasing mean MeDIP-seq signals for the low range of coupling factors.

Instead of transferring the identified normalization parameters to a computationally demanding Bayesian deconvolution process [27], the raw MeDIP-seq signals are weighted with respect to the estimated coupling factor dependent normalization parameters (see section 3.3). The main impact of this simplification is a significantly reduced run time for processing MeDIP-seq data by orders of magnitude. In the following, I demonstrate the performance of the MEDIPS procedure by comparing available and novel raw and normalized MeDIP-seq data to bisulphite sequencing derived methylation values. Moreover, the results of the MEDIPS procedure are compared to the results of the BATMAN software [27] with respect to normalization and runtime. In addition, raw and normalized MeDIP-seq derived promoter methylation is evaluated with respect to the effect on expression of downstream genes.

## Comparison to bisulphite sequencing data from sperm and to the results of BATMAN

In order to test the performance of the MEDIPS procedure, I processed MeDIP-seq data derived from a sperm sample published by Down et al. [27]. Additionally, I downloaded normalized methylation values provided by Down et al. [27] as well as benchmark methylation data derived from bisulphite-sequencing of another sperm sample generated by the human epigenome project (HEP) [30].

The analysis revealed that both normalization methods improved the poor correlation of raw data from MeDIP-seq and bisulphite sequencing from a Pearson correlation of 0.42 (Figure 3.10 a) to 0.83 (MEDIPS *ams* values, Figure 3.10 b) and 0.82 (BATMAN, Figure 3.10 c) respectively and that both methods have a high correlation of 0.92 (Figure 3.10 d).

## Runtime considerations

Until today, only the BATMAN software [27] was available for normalization of MeDIP-seq data. However, processing of MeDIP-seq data takes approximately three days for only one chromosome (i.e. human chromosome 1) on a modern-day server. As an example for the dire need for methods able to process MeDIP-seq data in a time-efficient way, researches in this field recently reported that they were not able to normalize their MeDIP-seq data in appropriate time [58]. In contrast, normalization of genome-wide MeDIP-seq data can now be achieved in only few hours by utilizing the novel software package MEDIPS (i.e. processing the full human genome on the same server as used for the BATMAN software).

## Comparison to bisulphite sequencing data from intestinal tissue

Genome wide methylation patterns are supposed to be unique for sperm samples and allow for distinguishing different cell-types [30, 85]. Furthermore, aberrant methylation can be associated with severe effects, for example the induction of cancer [47, 41]. In order to analyse methylation alterations in intestine tissue emerging during colon cancer development, *in-house* MeDIP-seq data of normal and tumor intestine tissues has been generated. Subsequently, I identified genomic regions showing differential methylation between normal and tumor tissues by applying MEDIPS. From the identified DMRs, 17

3 Modelling of MeDIP-seq data

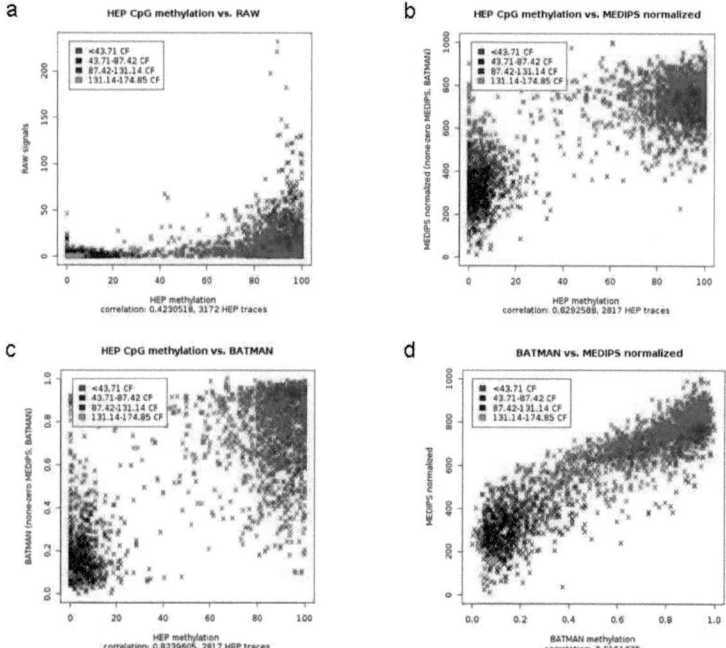

**Figure 3.10: Normalization of MeDIP-seq data.** I compared the normalization results of the MEDIPS package by processing available sperm MeDIP-seq data [27] against bisulphite sequencing derived methylation data of another sperm sample presented by the human epigenome project (HEP) [30]. Each data point within the figures represents a genomic region analyzed by bisulphite sequencing [30]. The colour code represents four quantiles of the calculated mean coupling factors (CpG densities) for theses regions. The figures show the comparison of raw (**a**), MEDIPS ams normalized (**b**), and BATMAN [27] normalized (**c**) MeDIP-seq data against bisulphite data from the HEP project [30]. **d** Comparison of MEDIPS normalized (*ams* values) against BATMAN [27] normalized MeDIP-seq data for the bisulphite sequenced genomic regions.

## 3 Modelling of MeDIP-seq data

| chr | start | stop | Correlation to RPM | Correlation to RMS |
|---|---|---|---|---|
| 3 | 370xxx | 370xxx | 0.999 | 0.999 |
| 6 | 133xxx | 133xxx | 0.893 | 0.941 |
| 17 | 753xxx | 753xxx | 0.927 | 0.973 |
| 12 | 133xxx | 133xxx | 0.911 | 0.987 |
| 5 | 134xxx | 134xxx | 0.928 | 0.924 |
| 10 | 131xxx | 131xxx | 0.913 | 0.919 |
| 3 | 148xxx | 148xxx | 0.994 | 0.994 |
| 12 | 101xxx | 101xxx | 0.852 | 0.882 |
| 2 | 193xxx | 193xxx | 0.982 | 0.963 |
| 6 | 163xxx | 163xxx | 0.255 | 0.255 |
| 1 | 356xxx | 356xxx | 0.969 | 0.944 |
| 4 | 107xxx | 107xxx | 0.965 | 0.943 |
| 5 | 388xxx | 388xxx | 0.847 | 0.875 |
| 14 | 100xxx | 100xxx | 0.850 | 0.904 |
| 20 | 480xxx | 480xxx | 0.644 | 0.659 |
| 8 | 115xxx | 115xxx | -0.385 | -0.437 |
| 20 | 610xxx | 610xxx | 0.967 | 0.969 |
|  |  |  | Median: 0.91 | Median: 0.94 |

**Table 3.3: Bisulphite validation of MeDIP-seq data from intestine cancer tissues of colon cancer patients.** Median correlation between bisulphite and MeDIP sequenicng data over all 17 tested genomic regions is 0.91 when RPM signals are considered and is improved to 0.94 when normalized RMS values are considered instead.

genomic regions were selected for an independent validation of methylation patterns by bisulphite sequencing.

Table 3.3 lists genomic regions (hg19) tested for methylation patterns in normal and tumor tissue of three different colon cancer patients. Tested genomic regions have an average length of 285 bp. For each tested region, correlation was calculated by comparing averaged bisulphite derived methylation values from normal and tumor tissues of the different patients against the correponding averaged MeDIP-seq derived RPM or normalized RMS values, respectively. Although median correlation over all 17 genomic regions is already 0.91 for RPM signals, the accordance between bisulphite and MeDIP sequencing data is improved to a median correlation of 0.94 when normalized RMS values are considered (see Table 3.3).

# 3 Modelling of MeDIP-seq data

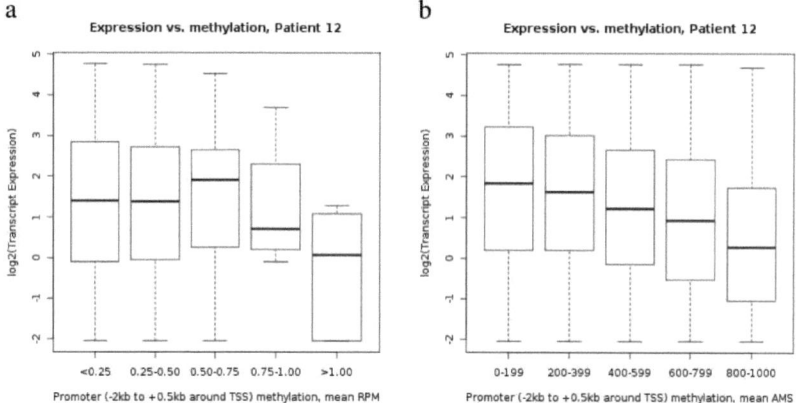

**Figure 3.11: Comparing promoter methylation and transcript expression.** Association of average intestine promoter methylation with transcript expression level. The boxes correspond to five regular quantiles of the RPM (or RMS, respectively) data range. **a** For each RPM bin, the transcript expression boxplot is reported. **b** Shows the same analysis with normalized AMS data on the X-axis.

### Comparison to gene expression patterns

It was previously observed that highly expressed genes show low promoter but elevated exon methylation level [85]. Promoter DNA methylation is expected to determine transcriptional repression of the downstream gene [80]. In order to test the dependency between promoter methylation and transcript expression, I have accessed available *in-house* RNA-Seq transcript expression data from normal intestine tissue of one colon cancer patient and compared expression to MeDIP-seq derived methylation levels of the same tissue and patient. Figure 3.11a shows that association between promoter methylation (X-axis) and transcript expression (Y-axis) does not reveal a general trend when averaged RPM signals are considered. In contrast, averaged normalized AMS values of promoters show improved negative correlation with the expression of downstream transcripts (see Figure 3.11b).

In summary, the impact of MeDIP-seq data normalization by the MEDIPS procedure is demonstrated with respect to various aspects. First, improved concordance of normalized MeDIP-seq and bisulphite sequencing data is shown for available sperm and for novel intestine tissue data. Moreover, the performance of MEDIPS is compared to the BAT-

MAN software with respect to normalization results and runtime. Finally, it is shown that negative correlation of promoter methylation and gene expression regulation is improved when MeDIP-seq data is processed and normalized by the MEDIPS procedure. Therefore, MEDIPS is the first standard pipeline for MeDIP-seq data analysis able to cope with the inherent complexity of MeDIP-seq data and out performs computation time of existing methods by orders of magnitude with similar performance.

## 3.4 Identification of differentially methylated regions (DMRs)

Identification of DMRs is essential for determining local differences in the methylation profiles of diverse biological samples. While there exist several methods for determining statistically significant enriched genomic regions from ChIP-on-Chip [59, 99, 17, 46] and ChIP-Seq experiments [45, 101, 67, 14, 88], the identification of differentially methylated regions from MeDIP-seq data remains an open problem. The main difference between the ChIP-Seq and MeDIP-seq approaches is that TFBSs are of short length (8-16bp) and therefore, ChIP-Seq specific methods intend to identify isolated short genomic regions of high short read enrichments. In contrast, CpGs are spread more widely along the chromosomes and are partly accumulated in CpG islands of length >300bp. Moreover, methylation alterations may occur only at few CpG locations and therefore, no sharp TFBSs like ChIP-Seq peaks are expected. Subsequently, in order to identify DMRs, comparatively longer genomic stretches have to be considered and methylation alterations have to be determined in a more sensitive way.

For identification of DMRs, there are two alternative approaches. First, it is of interest to specify pre-defined genomic regions of interest (ROIs) like CpG islands, promoters etc., and to specifically compare methylation patterns for these regions. Second, it is of interest to calculate differential methylation for genome wide frames of arbitrary length. However, in both cases I call any predefined genomic region as ROI. Here, I present a statistical approach for calculating differential methylation for any predefined ROIs, based on sequencing data from two different MeDIP treated samples (control and treatment) and with respect to an additional input sequencing data set (input).

# 3 Modelling of MeDIP-seq data

Let $C$, $T$, and $I$ be the genome vectors (see section 3.1) generated based on the sequencing data from control, treatment, and input sequencing data using an arbitrary bin size $b$ and let $ROI$ be a set of predefined ROIs, where $ROI = ROI_1, ..., ROI_i, ..., ROI_n$, and $n$ is the number of ROIs to be tested and the $ROI$s are of length $m_1, ..., m_n$. In the following, identification of DMRs is only supported for any $ROI_i$ of length $m_i \geq 5 \cdot b$. Therefore, each $ROI_i$ includes at least five genomic bins $b_{ij}$, where $b_{i1}, ..., b_{ij}, ..., b_{ik_i} \in ROI_i$ and $k_i = \lfloor \frac{m_i}{b} \rfloor$. For each $ROI_i$, mean $rpm$ (see subsection 3.3.1) and mean $rms$ (see subsection 3.3.4) values are calculated based on $C$ and $T$ as:

$$C.RPM_{ROI_i} = \frac{1}{k_i} \sum_{j=1}^{k_i} rpm(C.b_{i,j})$$

$$C.RMS_{ROI_i} = \frac{1}{k_i} \sum_{j=1}^{k_i} rms(C.b_{i,j})$$

$$T.RPM_{ROI_i} = \frac{1}{k_i} \sum_{j=1}^{k_i} rpm(T.b_{i,j})$$

$$T.RMS_{ROI_i} = \frac{1}{k_i} \sum_{j=1}^{k_i} rms(T.b_{i,j})$$

where $rpm(C.b_{i,j})$, $rms(C.b_{i,j})$, $rpm(T.b_{i,j})$, and $rms(T.b_{i,j})$ are the pre-calculated $rpm$ (see subsection 3.3.1) and $rms$ (see subsection 3.3.4) values of the genomic bins from the control and treatment samples. In addition, for each $ROI_i$, mean $rpm$ values are calculated based on $I$ as:

$$I.RPM_{ROI_i} = \frac{1}{k_i} \sum_{j=1}^{k_i} rpm(I.b_{i,j})$$

where $rpm(I.b_{i,j})$ are the pre-calculated $rpm$ (see subsection 3.3.1) values of the genomic bins from the Input sample.

# 3 Modelling of MeDIP-seq data

Based on the mean *rms* values of the control and of the treatment sample, for each $ROI_i$ the following ratios are calculated:

$$r.rms_{ROI_i} = \frac{C.RMS_{ROI_i}}{T.RMS_{ROI_i}}$$

In addition, by considering the mean *rpm* values of the control or of the treatment sample, respectively, the following ratios are calculated with respect to *rpm* values of the input sample:

$$r.rpm.C_{ROI_i} = \frac{C.RPM_{ROI_i}}{I.RPM_{ROI_i}}$$

$$r.rpm.T_{ROI_i} = \frac{T.RPM_{ROI_i}}{I.RPM_{ROI_i}}$$

Because local background sequencing signals are variable along the chromosomes due to differing DNA availability, a global background *rpm* signal threshold is estimated based on the distribution of all calculated $I.RPM_{ROI_i}$ values. This is done by defining a targeted quantile *qt* (e.g. $qt = 0.95$) and by identifying the $I.RPM_{ROI_i}$ value ($t$), where *qt*% of all $I.RPM_{ROI_i}$ values are $< t$. This estimated global minimal mean rpm threshold $t$ will serve as an additional parameter for selecting genomic regions that show a mean MeDIP-seq derived *rpm* signal of at least $t$ in the control or the treatment sample, respectively.

In addition, statistical testing is utilized in order to rate whether the obtained *rms* data series of the genomic bins within any $ROI_i$ significantly differ in the control sample compared to the treatment sample. As explained in section 2.2, for each $ROI_i$ it is tested, whether the *rms* values of the genomic bins $b_{i1}, ..., b_{ij}, ..., b_{ik_i} \in ROI_i$ of the control sample significantly differ from the *rms* values of the according genomic bins of the treatment sample. For this, the MEDIPS package (see section 3.5) utilizes the *t.test()* and *wilcox.test()* functions of the R environment (www.R-project.org) with default parameter settings (two-sided tests in both cases). Therefore, for each tested $ROI_i$, two p-values ($ROI.p.value.t_i$ and $ROI.p.value.w_i$) will be calculated and serve as a further level for discriminating between local methylation profiles.

## 3 Modelling of MeDIP-seq data

For identifying $ROI_i$'s that show differential methylation between the control and the treatment sample with respect to the input sample, based on the pre-calculated parameters, a filtering procedure is performed. The following filtering procedure also discriminates between increased methylation in the control sample compared to the treatment sample (control>treatment, a) and vice versa (treatment>control, b):

1. $ROI_i$'s where $C.RMS_{ROI_i} = T.RMS_{ROI_i} = 0$ are neglected,

2. $ROI_i$'s where $ROI.p.value.t_i > p$ and $ROI.p.value.w_i > p$ are neglected, where $p$ is any targeted level of significance (e.g. $p = 0.01$),

3. filtering for the ratio:

    a) $ROI_i$'s where $r.rms_{ROI_i} < h$ are neglected, where $h$ is an upper ratio threshold (e.g. $h = 1.33$),

    b) $ROI_i$'s where $r.rms_{ROI_i} > l$ are neglected, where $l$ is a lower ratio threshold (e.g. $l = 0.75$),

4. filtering for global input derived background signals:

    a) $ROI_i$'s where $C.RPM_{ROI_i} < t$ are neglected,

    b) $ROI_i$'s where $T.RPM_{ROI_i} < t$ are neglected,

5. filtering for local input derived background signals:

    a) $ROI_i$'s where $r.rpm.C_{ROI_i} < h$ are neglected,

    b) $ROI_i$'s where $r.rpm.T_{ROI_i} < h$ are neglected.

The remaining $ROI$'s are considered as candidate genomic regions where events of differential methylation can be deduced from the data in a sophisticated statistical way.

As mentioned above, a set of $ROI$'s can be defined as adjacent genome wide windows of an arbitrary constant length (e.g. a frame size $frame = 500$ bp). Moreover, neighboring $ROI$'s are allowed to overlap by any arbitrary constant length (e.g. by an overlap of $overlap = 250$ bp), where it must be guaranteed that $overlap \geq 0$ and $overlap \leq frame$. Whenever an $overlap > 0$ was defined, significant overlapping DMRs can be received. Therefore, the MEDIPS package (see section 3.5) merges identified overlapping DMRs into one supersized region.

## 3.5 MEDIPS software package

The MEDIPS software package was developed for analyzing data derived from methylated DNA immunoprecipitation (MeDIP) experiments followed by sequencing (MeDIP-seq). Nevertheless, functionalities like the *saturation analysis* (see subsection 3.2.1) may be applied to other types of sequencing data (e.g. ChIP-Seq).

MEDIPS incorporates all of the developed methods in the context of MeDIP-seq data analysis as presented in sections 3.1-3.4. Moreover, MEDIPS simplifies processing of MeDIP-seq data as it starts where the mapping tools stop and allows for exporting of the results for visualization in common genome browsers. MEDIPS is available as an R package, is suitable for any arbitrary genome available via BioConductors [33] annotation libraries [77] and provides comprehensive functionalities for accelerated processing of MeDIP-seq data.

The main features of the package are:

- calculating genome wide MeDIP-seq signal densities at an user specified resolution (see section 3.1),

- estimating the reproducibility for obtaining full genome methylation profiles with respect to the total number of given short reads and with respect of the size of the reference genome (see subsection 3.2.1),

- analyzing the coverage of genome wide DNA sequence pattern (e.g. CpGs) by the given reads (see subsection 3.2.2),

- calculating CpG enrichment factors as a quality control for the immuno-precipitation (see subsection 3.2.3),

- calculating genome wide sequence pattern densities (e.g. CpGs) at an user specified resolution (see subsection 3.3.2),

- plotting of calibration plots as a data quality check and for a visual inspection of the dependency between local sequence pattern (e.g. CpG) densities and MeDIP signals (see subsection 3.3.3)

## 3 Modelling of MeDIP-seq data

- normalization of MeDIP-seq data with respect to local sequence pattern (e.g. CpG) densities (see subsections 3.3.4 and 3.3.5),

- summarizing methylation values for genome wide windows of a specified length or for user supplied regions of interest (ROIs),

- calculating differentially methylated regions on raw or normalized data comparing two sets of MeDIP-seq data with respect to input-Seq data (see section 3.4),

- export *rpm* (see subsection 3.3.1) and normalized *rms* (see subsection 3.3.4) data for visualization in common genome browsers (e.g. the UCSC genome browser [86]).

- annotation of identified DMRs with respect to given annotation files containing genomic coordinates of e.g. promoter regions, exons, introns, CpG islands, etc.

A schematic diagram of a typical workflow for processing MeDIP-seq data by the MEDIPS software package is shown in Figure 3.12. The package comes along with a manual describing all steps of the workflow. It is recommended to have a look for software and manual updates at http://medips.molgen.mpg.de/ or at http://www.bioconductor.org/.

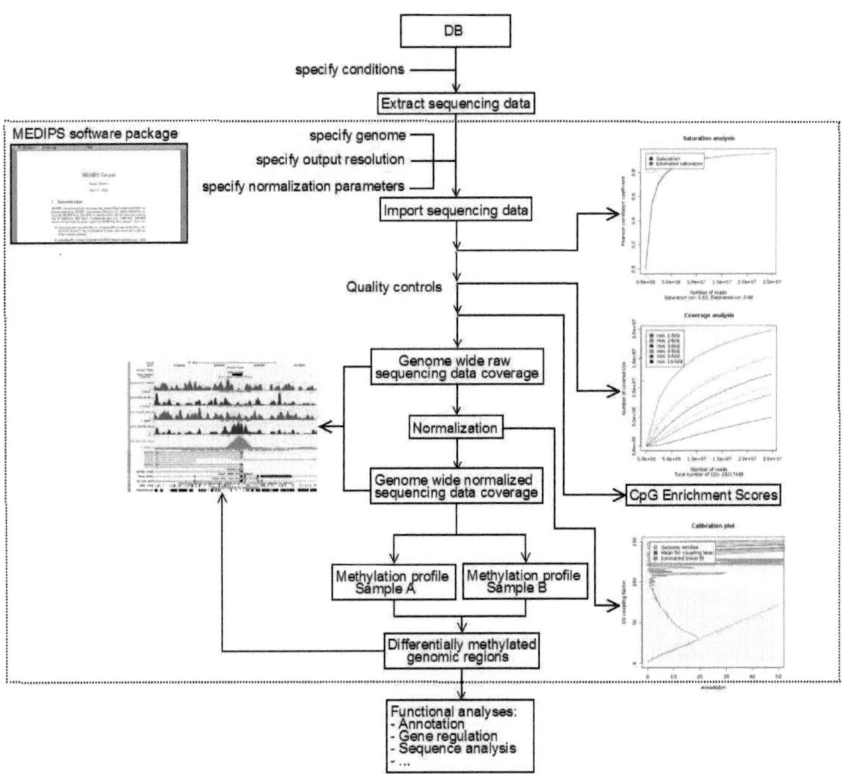

**Figure 3.12: Schematic diagram of the MEDIPS software package.** After having specified the reference genome, output resolution, and further normalization parameters, MEDIPS imports short read information, and subsequently, quality control metrics can be calculated. Genome wide short read coverages can be exported for visualization in common genome browsers in RPM format and after normalization in RMS format, respectively. Differential methylation is calculated by comparing methylation profiles of different samples. Identified DMRs can be exported or annotated for functional known genomic regions, like e.g. promoters, exons, introns or CpG islands.

# 4 Epigenetic dependencies during endodermal differentiation of human ES cells

An additional important aim of this study is the comparative analysis of genetic and epigenetic regulatory dependencies during differentiation of human embryonic stem cells along the endodermal lineage. Especially the coherences of genome-wide DNA methylation profiles, transcription factor binding sites (e.g. OCT4), altered gene expressions, and differentially methylated genomic regions emerging upon initiation of differentiation along the endodermal lineage will be evaluated. Finally, DNA methylation and gene expression alterations will be shown in detail for a reconstructed core OCT4 network controlling pluripotency in hESCs.

In order to analyse genetic and epigenetic dependencies during endodermal differentiation of human embryonic stem cells, it was necessary to derive definitive endoderm (DE) from hESCs and to measure the resulting gene expression and DNA-methylation profiles of both cell types. This has been done using the Illumina beadarray (see also section 2.1) and MeDIP-seq (see also sections 2.3 and 2.5) technologies.

Here, it is necessary to mention that these wet-lab experiments are not the direct content of this book and were performed by co-workers as a request for this work. Because detailed informations on the wet-lab experimental approaches are indispensable for interpretation of the results, section 4.2 describes the wet-lab experiments and protocols performed by the co-workers. However, all aspects of data analysis, cross-linking, and interpretation were performed exclusively by the author.

## 4.1 Identifying a core regulatory network of OCT4 controlling pluripotency

The TF OCT4 is known as a key regulator for maintaining pluripotency in the mammalian embryo [73, 81] and functional data on OCT4 regulatory action is available from heterogeneous sources: ChIP-on-chip experiments [13], promoter sequence analysis [93, 18, 87], and RNA interference [6] provide complementary pieces of information on OCT4 transcriptional dependencies (see also section 1.2). In order to identify a core OCT4 regulatory network in hESCs, I have performed [17] an integrated analysis of such high-throughput data along with promoter sequence analysis.

**Analysis of individual experimental methods**

I performed a re-analysis of available OCT4, SOX2 and NANOG ChIP-on-chip data from hESCs [13] including mapping of the 60 mer oligonucleotide probes of utilized DNA tiling arrays to an updated NCBI build (v36.1). Processing of uniquely-mapped probes includes background correction, normalization, fold-enrichment and peak identification and resulted in 308 potential OCT4 target genes.

Because protein-DNA binding events do not give information on the direction of the regulatory influence of the TF with respect to the transcription rate of its target genes, I complemented the results of the OCT4 ChIP-on-chip experiment with the results of the RNAi mediated OCT4 silencing in hESCs performed by Babaie et al. [6]. Identifier mapping of the different chip platforms (Agilent oligochips and cDNA microarrays) resulted in 10,065 genes that were represented as cDNA clones on the microarray and that had promoter regions covered by the Agilent tiling arrays. From the originally published 623 OCT4 target genes [13], 472 were also represented on the cDNA microarray. From the 1,104 genes that show significantly altered expression 72 hours after the OCT4 knock down, 40 genes (<4%) were also identified as direct OCT4 target genes.

In order to obtain an even more stringent set of OCT4 target genes, I searched the promoter sequences of the targets for the occurrence of known OCT4-related octamer and SOX-OCT joint motifs within a distance of 8 kb upstream of the respective TSSs.

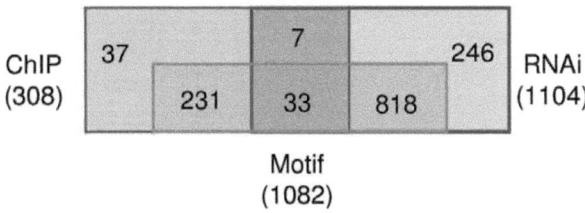

**Figure 4.1: Overlap of individual studies.** Overlap of the re-analyzed OCT4 ChIP-on-chip experiment, the OCT4 RNAi experiment and the motif mapping results with the octamer and sox-oct joint motifs.

Even though I neglect information on binding events caused by OCT4-DNA interactions mediated by unknown cofactors and heterodimer complexes, the results reflect confirmed functional circuitries dependent on direct OCT4 and SOX-OCT binding. Combination of the three approaches resulted in a set of 33 genes (see figure 4.1).

### OCT4 target genes

Among the 33 genes, several well-known targets of OCT4 can be found as well as genes whose regulatory interaction with OCT4 is less well-described. In general, OCT4 binds to and regulates diverse classes of genes encoding for example transcription factors (TGIF2, EOMES, FOXD3, GSC, TSC22D1, GATA6, OCT4, SOX2, NANOG, PAX6, CDX2, TCF4), transcriptional regulators (SSBP2), regulators of kinase, transferase, and catalytic activity (GAP43, TDGF1), members of the Wnt receptor signalling pathway (SFRP2, FRAT2, DKK1), and growth factors (FGF2, LEFTY2, TDGF1).

### OCT4 core regulatory network

The resulting OCT4 core regulatory network, also incorporating the information on direct target genes from the re-analyzed SOX2 (red lines) and NANOG (blue lines) ChIP-on-chip experiments, is shown in figure 4.2.

The network distinguishes genes that are suppressed (left side) from those that are activated (right side) by OCT4. Among the 33 genes a high fraction is annotated with

# 4 Epigenetic dependencies during endodermal differentiation of human ES cells

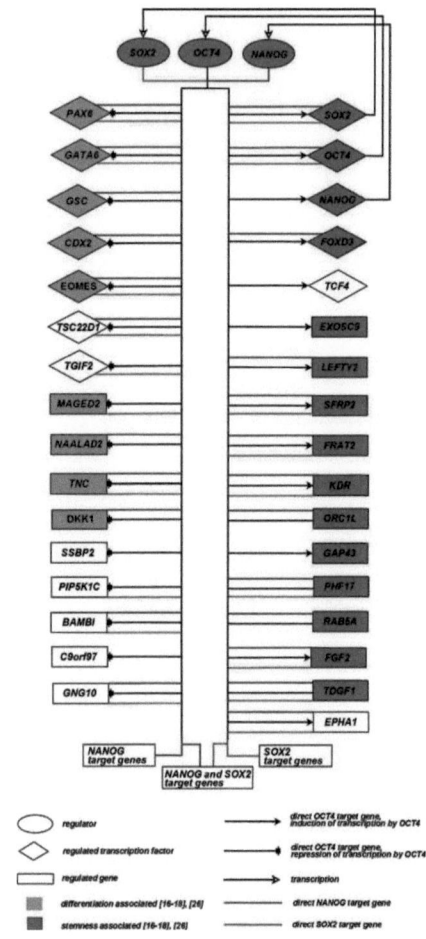

**Figure 4.2: OCT4 core regulatory network.** Core OCT4 transcriptional regulatory network identified by the integrative analysis of the re-analysed ChIP-on-Chip data, the OCT4 RNAi knock-down and the sequence-based octamer and sox-oct motif mapping. Green boxes represent genes associated with differentiation and red boxes indicate genes being specific for hESCs as annotated by several further public sources. For white boxed genes no detailed annotation about differentiation or stemness characteristics was found by literature research. The network also incorporates the information on direct target genes from the re-analyzed SOX2 (red lines) and NANOG (blue lines) ChIP-on-chip experiments.

transcription factor activity (GO:0003700, indicated as rhombuses). Furthermore, a classification in hESCs specific genes (red boxes) and genes that are associated with the process of differentiation (green boxes) was performed by accessing several further public sources [6, 2, 75, 5]. White boxed genes could not be annotated using these sources, but the information about up or down regulation after the OCT4 knock-down indicates, whether the respective gene is functional connected to the process of differentiation or to the maintenance of pluripotency.

An additional level of gene regulation has been added to this core OCT4 target network by further literature and database mining (see section 1.2). Figure 1.2 shows the core network extended by known up- and downstream target genes of the respective TFs as given by TRANSFAC [68] and by another published work [34].

By this integrative approach, I reconstructed the OCT4 dependent functional transcriptional regulatory network important in the analysis of human stem cell characteristics and cellular differentiation. Functional information is largely enriched using an overlay of different experimental results [17].

## 4.2 Experimental procedures and analysis methods

### Differentiation of human embryonic stem cells (hESCs) into definitive endoderm (DE)

Upon treatment with Activin A (100ng/ml) for 5 days, undifferentiated human ES cells (H1, passage 53) changed morphology from typical, defined, tight colonies (figure 4.3 a) into less dense, flatter cells (figure 4.3 b) [24].

In order to confirm the differentiation into definitive endoderm (DE), we detected the expression SOX17 using immunostaining (figure 4.3 c) and investigated lineage-specific genes expression patterns by real-time RT-PCR (figure 4.3 d). After 5 days of Activin A treatment, the majority of the cells were devoid of the pluripotent marker OCT4 but however showed expression of the transcription factors SOX17 and FOXA2 which are markers of DE.

# 4 Epigenetic dependencies during endodermal differentiation of human ES cells

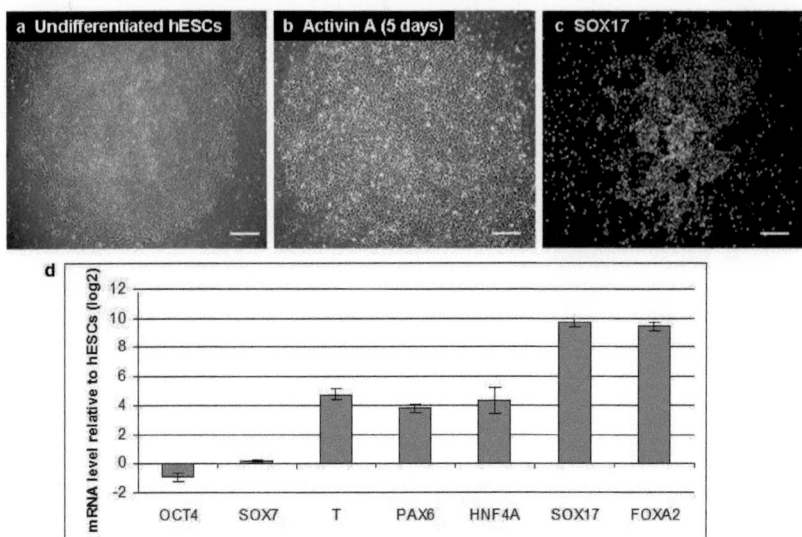

**Figure 4.3: Derivation of definitive endoderm from human ES cells. a** Phase contrast image of undifferentiated human ES cells and b cells after 5 days of Activin A treatment. **c** Immunofluorescence labeling of differentiated cells demonstrating SOX17 expression. Scale bar = $100\mu m$. **d** Effect of Activin A treatment on the gene expression of selected genes during differentiation of human ES cells. The ratios represent the mean of two independent biological replicates. Bars indicate standard errors (SE) between the biological replicates.

Importantly, there was low level expression of the transcription factor SOX7 (expressed in primitive endoderm but not in DE). This implies that the induction of SOX17 and FOXA2 expression was not a result of differentiation into primitive endoderm. PAX6 expression is detectable, demonstrating the presence of some ectodermal cells. Moreover, Brachyury (T) expression was also detected, which might imply a transition through the primitive streak stage of development. Furthermore, HNF4A is up-regulated therefore indicative, of early hepatic-like characteristics of the Activin A treated cells.

**Library preparation**

Genomic DNA was sonicated for 2 h as described previously [78] to a size range of 100-400 bp. Fragmented DNA was purified using QIAquick spin columns and buffer QG (Qiagen) according to the manufacturer's protocol. 5 μg of fragmented DNA were subjected to single end library preparation using the genomic DNA sample prep kit (#FC-102-1002, Illumina) according to the manufacturer's instructions with the following modifications: 2.4 times increased amount of enzymes were used for end-repair and A-tailing.

End repair was performed in the presence of 0,25 mM dNTPs Mix in a total volume of 317 μl, A-tailing was performed in a total volume of 88 μl. Adapters were ligated to the DNA fragments using 29 μl of 'Adapter oligo mix' and 2 times excess concentration of ligase in a total reaction volume of 98 μl. The sequencing libraries were subjected to immunoprecipitation (see below).

The amplification was performed after immunoprecipitation prior to gel-size selection. 20 % of the immunoprecipitated DNA or 40 ng of sheared DNA (input) were amplified using 6 amplification cycles in a total volume of 30 μl. Amplified libraries were size selected on a 2 % agarose gel to fragments of 150-400 bp (corresponding to insert sizes of 80-330 bp). Libraries were quantified on a Qubit fluorometer using the QuantIt dsHS Assay Kit (Invitrogen).

**Immunoprecipitation of methylated DNA (MeDIP)**

MeDIP was adapted from a previously published protocol [103]. 10 μl of monoclonal antibody against 5-methylcytosine (#BI-MECY, Eurogentec) were coupled overnight with 40 μl Dynabeads M-280 sheep anti-mouse IgG (Invitrogen) in 500 μl 0.5% BSA/PBS, washed two times with 0.5% BSA/PBS and once with IP-buffer (10 mM sodium phosphate (pH7.0), 140 mM NaCl, 0.25 % Triton X100).

The sequencing libraries were denatured for 1 min at 95 °C. 4 μg of library was immunoprecipitated for 4 h at 4 °C with the 5-methylcytosine antibody coupled to Dynabeads in 230 μl IP-buffer, then washed three times with 700 μl IP-buffer. The beads were treated with 50 mM Tris-HCl, pH 8.0; 10 mM EDTA, 1 % SDS for 15 min at 65 °C and collected using a magnetic rack. The supernatant containing the methylated DNA (200 μl) was diluted with 200 μl 10 mM Tris pH 8,0, 1 mM EDTA, treated with proteinase K (0.2 μg/μl) for 2 h at 55 °C, followed by phenol-chloroform-extraction and ethanol precipitation. The precipitated DNA was resuspended in 20 μl of 10 mM Tris pH 8.5.

**Illumina Genome Analyzer sequencing**

After library quantification at a Qubit (Invitrogen) a 10 nmol stock solution of the amplified library was created. 12 pM of the stock solution were loaded onto the channels of a 1.4 mm flow cell and cluster amplification was performed. Sequencing-by-synthesis was performed on a Illumina Genome Analyser (GAIIx). After quality control of the first base incorporation (signal intensities, cluster density) the run was started. All MeDIP and input samples were subjected to 36 b single read sequencing run.

**Gene expression beadarray experiments**

Biotin-labeled cRNA was produced by means of a linear amplification kit (Ambion, Austin, TX, USA) using 500ng of quality-checked DNase-free total RNA as input. Chip hybridisations, washing, Cy3-streptavidin staining, and scanning were performed on an Illumina BeadStation 500 platform (Illumina, San Diego, CA, USA) using reagents and following protocols supplied by the manufacturer. cRNA samples were hybridised on Illumina human-8 BeadChips. We hybridised undifferentiated and Activin A-treated (definitive endoderm-differentiated) H1 cell line (passage 53) samples in biological triplicates.

## 4.3 MeDIP-seq quality control metrics

The raw sequencing data processing was done using Illuminas proprietary image analysis and base calling pipeline version 1.4. After mapping [57] of the generated reads against the human genome hg19 build downloaded from UCSC [86] (http://genome.ucsc.edu/), we obtained ~25,9 million unique high quality (MAQ [57] quality score ≥10) mapping hits for pluripotent hESCs and ~32,6 million for DE. Additionally, we obtained ~22,6 million unique high quality mapping hits from input samples of both conditions.

Based on the high quality mapping hits of the generated short reads from hESCs, DE, and input, we first performed saturation analyses resulting in genome-wide coverage saturation of 0.94 for hESCs and 0.96 for DE (see figures 4.4 a and b).

Because the constellation of DNA-fragments that have to be sequenced is much higher for input samples than for immunoprecipitated samples, the estimated saturation for the

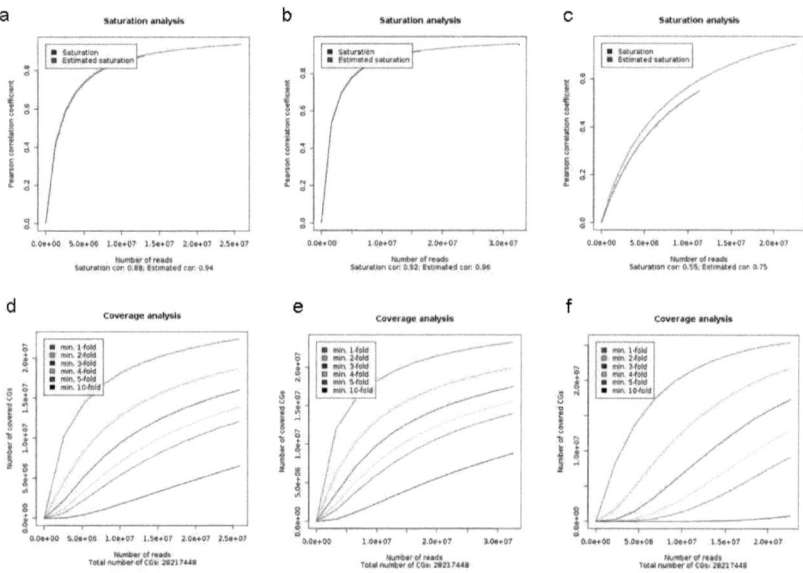

**Figure 4.4: Saturation and CpG coverage analysis.** The figure shows the results of the saturation analysis (see subsection 3.2.1) of the MEDIPS package analyzing MeDIP-seq data from hESCs (**a**), DE (**b**), and from the input samples (**c**). Additionally, the figure shows the results of MEDIPS coverage analysis (see subsection 3.2.2) for the MeDIP-seq data from hESCs (**d**), DE (**e**), and from the input samples (**f**).

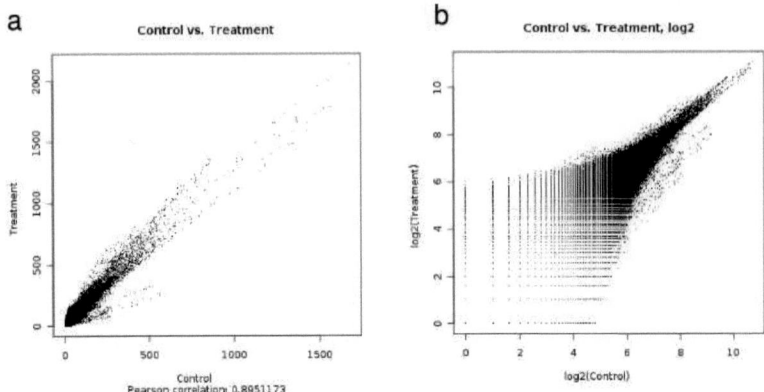

**Figure 4.5: Scatter plots comparing raw MeDIP-seq data from hESCs against DE.** The figure shows scatter plots comparing raw MeDIP-seq data sampled at genome wide 50bp bins for hESCs (control) and DE (treatment) in normal (**a**) and log2 (**b**) scale.

input sequences is lower (0.75) (see figures 4.4 c). Coverage analysis shows a good CpG coverage saturation of the ~28.2 million CpGs of the human genome. In the hESCs sample 22.4 million CpGs (79%), in the DE sample 23,2 million CpGs (82%), and in the input sample 25.4 million CpGs (90%) were covered at least once (see figures 4.4 d-f).

The genome wide Pearson correlation obtained when comparing MeDIP-seq data from the hESCs and DE samples is 0.9 (see figures 4.5 a and b).

Moreover, we tested the enrichment of CpG-rich short reads derived from the immunoprecipitation step and found a relative enrichment for CpG rich short reads from the hESC sample (2.11) and DE sample (2.59) compared to the reference genome, whereas, as expected, the relative CpG enrichment is close to one (1.16) for the combined input samples (see also Table 3.2).

Finally, the calibration curves clearly reveal the dependency between increasing MeDIP-seq signals and increasing local total CpG densities for the hESCs and DE samples resulting from immunoprecipitation, but not for the input sample (see figures 4.6 a-c).

4 Epigenetic dependencies during endodermal differentiation of human ES cells

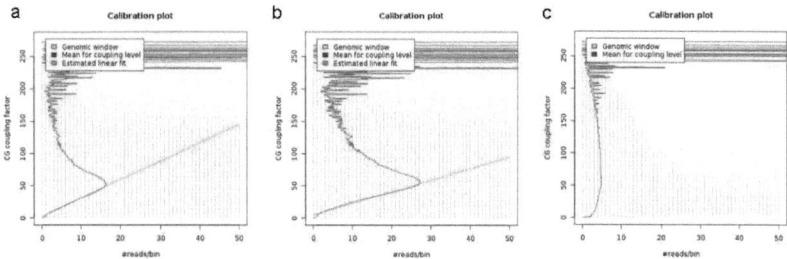

**Figure 4.6: Calibration plots.** The figure shows the calibration plots generated by the MEDIPS package after processing of the full-genome MeDIP-seq data from hESCs (**a**), DE (**b**), and from the input samples (**c**). Each data point within the calibration plot represents a 50bp bin. The calibration curve (red line) represents the dependency between MeDIP-seq signals and local CpG densities over the full range of coupling factors. The green lines in Figs. a and b show the estimated normalization curve calculated from the lower lever of coupling factors. For input samples (c), no CpG density dependent normalization was performed. The plots show the full range of coupling factors but are limited on the signal range (x-axis) showing only genomic bins associated to a maximum signal of 50 reads per bin.

## 4.4 Comparing MeDIP-seq and WGSBS derived methylation profiles

Recently, Lister et al. [64] presented full genome DNA methylation of hESCs at base resolution generated from 1.16 billion short reads of a whole-genome shotgun bisulphite sequencing (WGSBS) approach. Moreover, Lister et al. [64] showed that 25% of all methylated cytosines in hESCs exist in a non-CpG context. Although MeDIP derived methylation signals are not at a base resolution level, we were interested in comparing mean MeDIP-seq and mean WGSBS methylation values for defined regions of interest.

For all Ensembl [10] transcript promoters (-1kb to +0.5kb around their TSSs) of chromosome 1 that contain MeDIP-seq derived short-reads or WGSBS data (10,633), we calculated mean WGSBS derived CpG methylation values and compared them against according mean un-normalized (*rpm*, see subsection 3.3.1) MeDIP-seq values from hESCs.

The scatter plot in figure 4.7 a shows the resulting low correlation of 0.33. Figure 4.7 b shows that the low correlation is increased to 0.72 by normalizing the rpm MeDIP-seq signals into absolute methylation signals (ams) using MEDIPS. For CpG islands [97], the

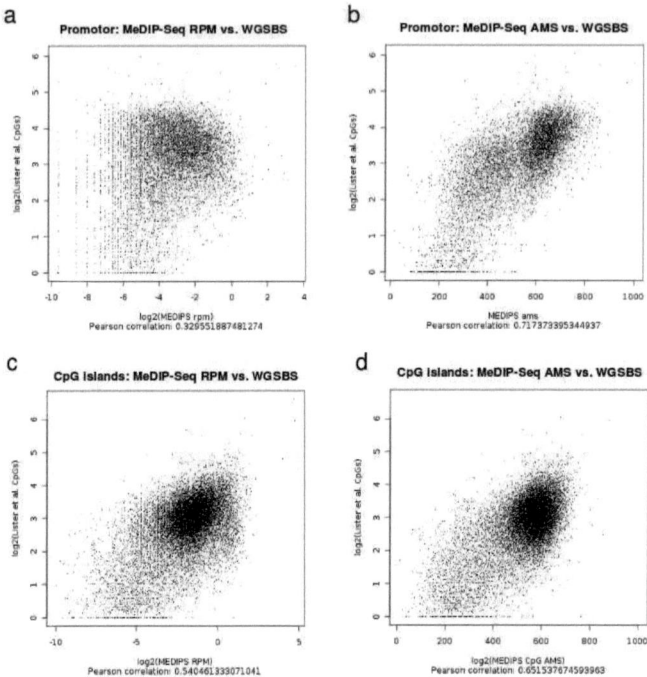

**Figure 4.7: Promoter and CpG islands methylation and comparison to WGSBS.** For 10,637 Ensembl [10] transcript promoters of chromosome 1, the mean WGSBS and *rpm* MeDIP-seq signals show a correlation of 0.33 (**a**). The WGSBS vs. MeDIP-seq correlation is increased to 0.71 after MEDIPS normalization of the MeDIP-seq signals into absolute methylation signals (*ams*) (**b**). For CpG islands, the correlation between mean *rpm* MeDIP-seq and mean WGSBS values is 0.59 (**c**) and is increased to 0.65 after MEDIPS normalization of the MeDIP-seq signals into absolute methylation signals (*ams*) (**d**).

correlation between mean *rpm* MeDIP-seq and mean WGSBS values is 0.54 (see figure 4.7 c) and is increased to 0.65 when considering MEDIPS normalized *ams* values (see figure 4.7 d).

## 4.5 Promoter methylation

We have, in particular, analyzed CpG density and methylation distributions in proximal promoter sequences (-1kb to +0.5kb around the TSSs) of 96,016 Ensembl [10] transcripts. Figure 4.8 a shows the well known bimodal CpG density distribution present in human

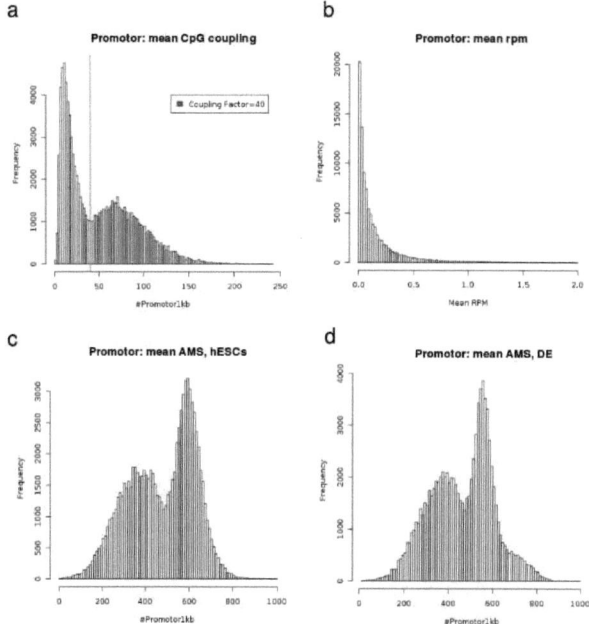

**Figure 4.8: Promoter CpG density and methylation distributions.** DNA sequences underlying human promoters [10] (-1kb to + 0.5kb around their TSSs) show a bimodal distribution of CpG densities (calculated as means of CpG coupling factors, **a**). By setting the coupling factor = 40, we define a threshold for discriminating between LCPs and HCPs. The bimodal methylation distribution present within promoters of human transcript cannot be revealed by accessing un-normalized MeDIP-seq data (*rpm*, see subsection 3.3.1) (**b**). MEDIPS normalized absolute methylation scores (*ams*, see subsection 3.3.5) reveal the bimodal promoter methylation distributions present in hESCs (**c**) and in DE (**d**).

promoters. By visual inspection of the plot, we define the coupling factor = 40 as threshold for discriminating between low CpG density (LCP, 48,021 transcripts) and high CpG density (HCP, 47,995 transcripts) promoters.

Whereas bimodal promoter methylation is not obvious when considering non-normalized *rpm* (see subsection 3.3.1) MeDIP-seq signals from hESCs (see figure 4.8 b), MEDIPS normalized *ams* (see subsection 3.3.5) MeDIP-seq values reveal the bimodal promoter methylation distribution present in hESCs (see figure 4.8 c) and in DE (see figure 4.8 d).

For hESCs, a large fraction of LCPs (22,104, 46%) is highly methylated (mean *ams* ≥

600) whereas only 3,488 (7%) LCPs show low methylation levels (mean $ams \leq 400$). For HCPs, this observation is reversed: 33,196 (69%) HCPs are lowly methylated, whereas only 189 (< 1%) HCPs are highly methylated. For DE, a similar trend was observed (data not shown).

## 4.6 Methylation patterns of transcription factor binding sites

We tested the ChIP-Seq derived transcription factor binding sites (TFBSs) of six transcription factors (TFs) as presented by Lister et al. [64] for their mean CpG densities and absolute methylation values (ams) in hESCs.

Interestingly, OCT4 TFBSs show bimodal CpG density and methylation distributions (see figures 4.9 a and b). In addition to OCT4, the binding sites of KLF4, and TAF1 show bimodal CpG density and absolute methylation signal distributions (see figures 4.9 c-f) but in contrast to OCT4 and KLF4 binding sites and in contrast to promoters, the majority of TAF1 binding sites are associated with high CpG densities and low methylation values.

Another class of transcription factors are NANOG, SOX2 and p300 as their binding sites cannot be distinguished into two groups of CpG densities or absolute methylation signals (see figure 4.9 g-l).

## 4.7 Identification of differentially methylated regions (DMRs)

Based on the MeDIP-seq data from hESCs, DE and input, respectively, we calculated the short read coverage (extend value=400) at genome wide 50bp bins using MEDIPS (see also section 3.1). In order to identify differentially methylated regions, MEDIPS calculates mean *rpm* (for hESCs, DE and input) and mean *rms* (for hESCs and DE, only) values for overlapping genome wide 500bp windows where neighboring windows overlap by 250bp (see also section 3.4).

Figure 4.10 shows histograms of mean *rpm* signals of the processed genomic windows for the input (blue curve), hESCs (control, red curve), and for the DE (treatment, green

4 Epigenetic dependencies during endodermal differentiation of human ES cells

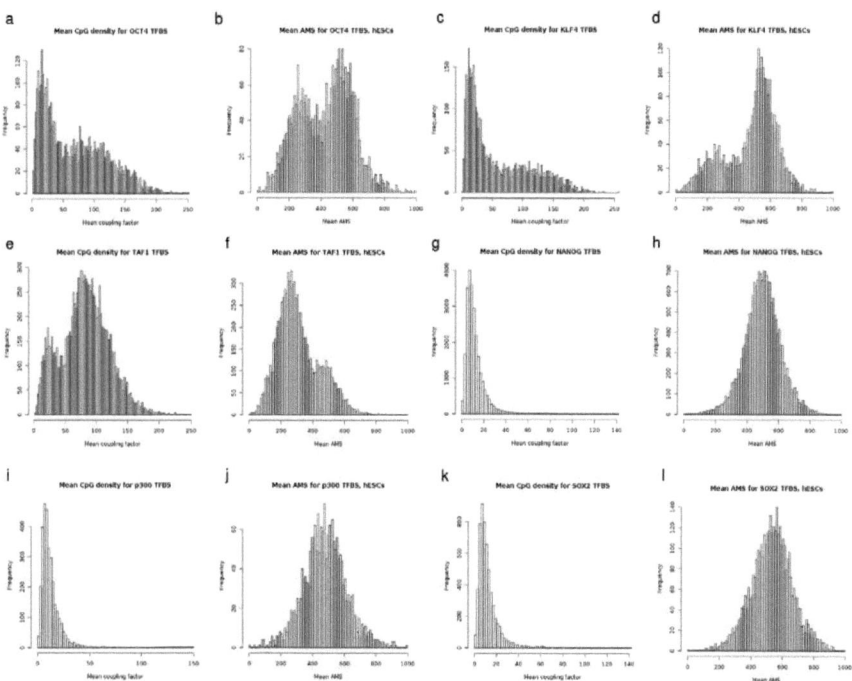

**Figure 4.9: CpG densities and absolute methylation scores of TFBSs.** OCT4 binding sites [64] show bimodal CpG density (**a**) and absolute methylation score (*ams*, see subsection 3.3.5) (**b**) distributions in hESCs. In addition to OCT4 binding sites, bimodal CpG density and absolute methylation score distributions can be observed for KLF4 and TAF1 binding sites [64] (**c-f**). No bimodal CpG density and absolute methylation score distributions can be observed for NANOG, p300 and SOX2 binding sites [64] (**g-l**).

## 4 Epigenetic dependencies during endodermal differentiation of human ES cells

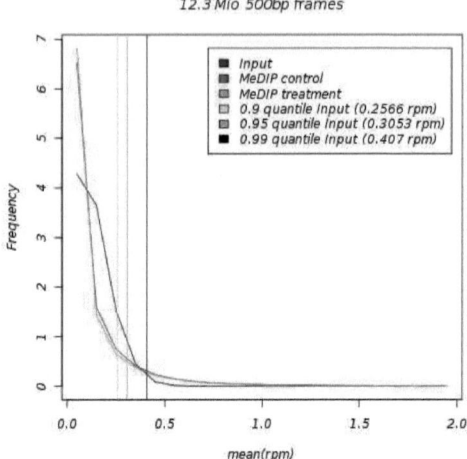

**Figure 4.10: Global rpm signal distributions.** The figure illustrates histograms for the mean *rpm* values of all genome-wide overlapping 500bp windows for the hESCs (control, red curve), DE (treatment, green curve), and input (blue line) samples. The grey lines indicate three possible global *rpm* thresholds obtained by setting $qt = 0.9$, $qt = 0.95$, and $qt = 0.99$ (grey lines).

curve) sample. In order to estimate a minimal global background signal threshold, the distribution of the mean *rpm* values derived from the input sample is examined. For this, MEDIPS calculates the 0.9 quantile ($qt$) of these pre-calculated mean *rpm* input values (see also section 3.4). By this approach, a *rpm* threshold of $t = 0.2566$ is identified. Therefore, 90% of all mean *rpm* input signals are $\leq t$. Figure 4.10 indicates three alternative *rpm* thresholds identified by the parameter settings $qt = 0.9$, $qt = 0.95$, and $qt = 0.99$ (grey lines).

Additionally, we calculated p-values by comparing the *rms* signal distributions of the 50bp bins of the hESCs and DE samples within each of the 500bp windows (see also section 3.4). Finally, de-methylation events (hESCs>DE) were identified by filtering for windows associated with a p-value $\leq 0.001$, with a mean hESCs *rpm* value $\geq t$ (defined by $qt = 0.9$), with a mean local *rpm* hESCs/input ratio $\geq 1.333333$, and with a mean *rms* hESCs/DE ratio $\geq 1.333333$. *De-novo* methylation events (DE>hESCs) were identified by filtering for windows associated with a p-value $\leq 0.001$, with a mean DE *rpm* value value $\geq t$ (defined by $qt = 0.9$), with a mean local *rpm* DE/input ratio $\geq 1.333333$, and

*4 Epigenetic dependencies during endodermal differentiation of human ES cells*

with a mean *rms* hESCs/DE ratio $\leq 0.75$ (see also section 3.4).

Because we have executed the according *MEDIPS.diffMethyl()* function of the MEDIPS package for overlapping 500bp windows, we partly received overlapping significant frames. Therefore, we finally merged overlapping regions into one super sized region using the *MEDIPS.mergeFrames()* function of the MEDIPS package.

Finally, we identified 62,142 distinct genome wide regions that become de-methylated during the differentiation of hESCs into DE. On the other hand, we identified 10,435 genomic regions where *de-novo* methylation occurs during early differentiation along the endodermal linage.

The comparatively higher number of de-methylated regions compared to *de-novo* methylated regions emphasizes the important role of de-methylation during embryonic differentiation. As a comparison, Lister et al. [64] identified approx. 6 million cytosines with higher levels of methylation in hESCs compared to differentiated fetal lung fibroblasts and only 124,162 thousand cytosines with higher levels of methylation in fetal lung fibroblasts compared to hESCs. From the 491 regions that are hypo-methylated in hESCs compared to fetal lung fibroblasts [64], we also identified 62 regions (13%) where a *de-novo* methylation event occurs during the differentiation into DE and only 5 regions (1%) that appear more methylated in hESCs compared to DE.

## 4.8 Genome wide distribution of DMRs

The heatmap in figure 4.11 a shows mean *rpm* MeDIP-seq values for the three biological replicates of hESCs, DE, and of the input samples for a subset of the identified DMRs (selected by the highest variance over the samples). This clustering approach clearly separates the hESCs, DE, and input samples into distinct groups. Additionally, the heatmap contains scaled CpG coupling factors of the DMRs.

Interestingly, DMRs that become de-methylated during the differentiation of hESCs are associated with low CpG densities and DMRs that become *de-novo* methylated are associated with higher CpG densities (see figure 4.11 a). In addition to this observation, we calculated CpG observed/expected [32] ratios as a measure for CpG density with respect to the amount of cytosines present in both strands of the DNA for both sets of DMRs, separately. Whereas the majority of de-methylated regions are associated with very low CpG obs/exp ratios, *de-novo* methylated regions tend to be associated with higher CpG obs/exp ratios indicating higher densities of CpGs (see figure 4.11 b).

For the identified de- (figure 4.11 c) and *de-novo* (figure 4.11 d) methylated regions, we tested, if they overlap with HCPs (high CpG promoters) or LCPs (low CpG promoters), (here, we tested an overlap with the -2kb to +0.5kb regions around the TSSs), CpG islands [97], CpG island shores, exons, and introns, or if they are located intergenic.

Interestingly, a higher percentage of *de-novo* methylated regions overlap with promoters (17.23%) or CpG Islands (37.15%) compared to the percentage of de-methylated regions (6.09% and 8.85%, respectively). We observed that <1% of all de-methylation events occur within HCPs, whereas 12.33% of all *de-novo* methylated regions overlap with HCPs. The percentage of de-methylated regions that overlap with introns is considerably higher (56.28%) compared to the percentage of *de-novo* methylated regions (31.43%).

In total, an overwhelming 78.53% of all genome-wide de-methylation events can be associated with transcript bodies or proximal promoters associated with 12,930 unique Ensembl [10] gene names (including miRNAs and others) whereas 53% of all *de-novo* methylation events can be associated with gene regions or proximal promoters of 4,787 unique Ensembl genes.

## 4.9 Differential methylation at TFBSs

We have tested the TFBSs of six transcription factors in hESCs as published by Lister et al. [64] for overlaps with regions identified as differentially methylated during endodermal differentiation of hESCs. In total, DMRs are not significantly enriched for any of the sets of TFBSs.

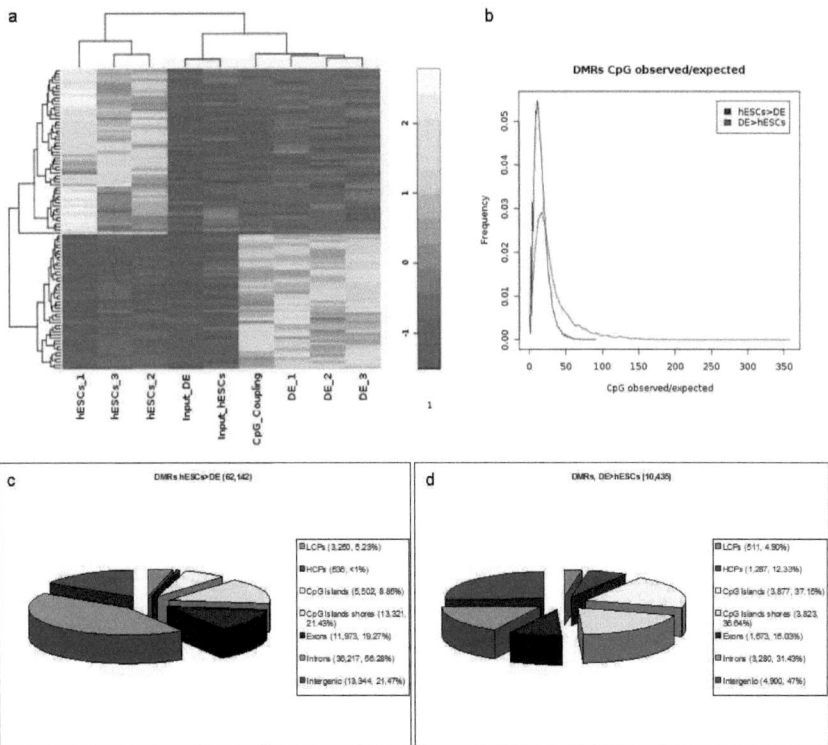

Figure 4.11: Differentially methylated regions (DMRs). a Heatmap of 100 DMRs including mean raw (rpm) MeDIP-seq signals for the three biological replicates of hESCs, and DE cells, the input sample from hESCs, the input sample from DE, and the mean CpG coupling factors. Differential methylation was calculated based on the pooled sets for hESCs, DE and input. b Distributions of CpG observed/expected [32] ratios for de-methylated regions (hESCs>DE) and *de-novo* methylated regions (DE>hESCs). The identified de-methylated (c) and *de-novo* (d) methylated regions were annotated for Ensembl [10] transcript promoters (-2kb to +0.5kb regions around their TSSs; divided into LCPs and HCPs), CpG islands [97] and their shores (-0.5kb form the start or +0.5kb from the end of a CpG island), exons, introns, and intergenic regions (no overlap with promoters and transcript bodies). Regions can be associated to more than one annotation (e.g. exon and CpG island).

4 Epigenetic dependencies during endodermal differentiation of human ES cells

| TF | #TFBS [64] | Ov. DMRs | Ov. DMRs (%) | Ov. hESCs>DE | Ov. DE>hESCs |
|---|---|---|---|---|---|
| OCT4 | 3,889 | 144 | **3.70** | 14 | 130 |
| KLF4 | 3,794 | 128 | **3.37** | 48 | 80 |
| TAF1 | 12,362 | 320 | **2.59** | 25 | 295 |
| SOX2 | 5,682 | 60 | **1.06** | 19 | 41 |
| NANOG | 25,071 | 321 | **1.28** | 117 | 204 |
| p300 | 3,093 | 35 | **1.13** | 4 | 31 |

**Table 4.1: Overlaps of TFBSs and DMRs.** For the six TFs [64] OCT4, KLF4, TAF1, SOX2, p300, and NANOG, the table lists the number of TFBSs that overlap with the identified DMRs (separated into de- and *de-novo* methylated regions). Ov. DMRs=number of TFBSs that overlap with DMRs; Ov. hESCs>DE=number of TFBSs that overlap with events of de-methylation; Ov. DE>hESCs=number of TFBSs that overlap with events of *de-novo* methylation.

However, de- and *de-novo* methylation occur within genomic regions identified as binding events of the TFs. For example, from the 3,889 OCT4 binding sites [64], there are 130 regions that become *de-novo* and only 14 regions that become de-methylated. Interestingly, although there are in total six times more DMRs that become de-methylated than *de-novo* methylated, the majority of DMRs that overlap with the TFBSs are associated with *de-novo* methylation for all six TFs (see Table 4.1).

Binding regions of the class of TFs that show bimodal methylation distributions (these are OCT4, KLF4, and TAF1), overlap more than twice as much with DMRs than TFBSs targeted by NANOG, SOX2, and p300.

## 4.10 Enrichment analysis associates de-methylation events to functional histone modifications

In order to further examine the identified DMRs, we have performed overrepresentation analyses for the de- and *de-novo* methylated regions separately, using the statistical analysis software EpiGRAPH [11]. Regions that become de-methylated during the differentiation of hESCs into definitive endoderm are among others significantly enriched for several distinct sequence patterns, known gene and transcript bodies [10], RefSeq gene bodies [84], predicted miRNAs [56], conserved regions [86], and for regions associated with selected histone modification events [7].

# 4 Epigenetic dependencies during endodermal differentiation of human ES cells

| Histone modification | Gene regulation | hESCs>DE | DE>hESCs |
|---|---|---|---|
| H2BK5me1 | + | 1 | 0 |
| H3K27me1 | + | 1 | -1 |
| H3K36me3 | + | 1 | 0 |
| H3K4me1 | + | 1 | 1 |
| H3K4me2 | + | 1 | 0 |
| H3K9me1 | + | 1 | 0 |
| H3R2me1 | + | 1 | 0 |
| H4K20me1 | + | 1 | 0 |
| PolII | + | 1 | 0 |
| H3K27me3 | - | -1 | 0 |
| H3K9me3 | - | -1 | 0 |
| H3K9me2 | - | -1 | 0 |
| H3K27me2 | - | 0 | -1 |
| H2A_Z | controversial | -1 | 0 |

**Table 4.2: Overrepresented histone modifications in DMRs.** The table shows all histone modifications [7] that are highly significantly (Bonferroni corrected) enriched (or underrepresented, respectively) within the identified de- and *de-novo* methylated regions. Statistical analysis was performed using EpiGRAPH [11] (here, EpiGRAPHSs *overlapRegionsCount* annotation was considered). Table legend: + associated with gene activation [7]; - associated with silent genes [7]; controversial no distinct effect [7]; 0 not associated to DMRs; 1 over-represented in DMRs; -1 underrepresented in DMRs.

Table 4.10 shows all histone modifications that are highly significantly (Bonferroni corrected) enriched (or underrepresented, respectively) for the identified de- and *de-novo* methylated regions (here, EpiGRAPHSs [11] *overlapRegionsCount* annotation was considered). DNA de-methylation is highly enriched for high signals of the gene activating monomethylations H3K27me1, H3K9me1, H4K20me1, and H2BK5me1 [7]. Moreover, de-methylation events are enriched for high levels of H3K4me1 and H3K4me2 which are positively correlated with transcriptional levels but occur in regions more downstream of TSSs, for high levels of H3K36me3 which were identified sharply after TSSs of active genes, for H3R2me1 which is modestly correlated with gene activation although no enrichment of H3R2me1 was found in the promoters of active genes over silent genes and for Pol II islands which are again positive correlated with gene expression [7].

On the other hand, events of DNA de-methylation are strongly underrepresented in regions connected to higher signals of H3K27me3, H3K9me3, and H3K9me2 which are connected to gene silencing [7]. Furthermore, de-methylation events are underrepresented in regions showing high levels of H2A.Z, whose effect to gene expression is controversial because its association with promoter regions is correlated with gene activity but its association within gene-bodies is correlated with gene silencing [7].

As already stated previously, the identified DMRs are not enriched for any set of the TFBSs of the six TFs analyzed by Lister et al [64], and moreover, EpiGRAPHs [11] enrichment analysis (performed for the TFBSs provided as custom annotations) shows that NANOG binding sites are even underrepresented in DNA de-methylation events. Interestingly, CpG islands are also no enriched targets of DNA de-methylation during endodermal differentiation of hESCs.

In contrast, *de-novo* DNA methylation events are strongly enriched in CpG islands. Moreover, *de-novo* methylation is predominantly present in Ensembl [10] transcript promoters (-2kb to +0.5kb around their TSSs) and in TFBSs conserved in human, mouse and rat [86]. For histone modifications, events of *de-novo* methylation are only enriched in regions associated to the gene activating monomethylation H3K4me1.

## 4.11 Differential methylation and gene expression alterations

In order to analyze the interplay between DNA-methylation and gene expression changes, we performed microarray-based gene expression analysis of hESCs and derived DE cells. Raw gene expression data was obtained employing the manufacturer's software BeadStudio 3.0.19.0. Subsequently, raw data was imported into the Bioconductor environment [33] and quantile normalization was performed using the beadarray package [29]. Figures 4.12 a and b show box plots of raw and normalized data. In order to test for global gene expression similarities within biological replicates and between different treatments, pairwise Pearson correlation coefficients were calculated for all samples. Correlations within the groups are all >0.99 and correlations between the groups are from 0.92 to 0.93 (see figure 4.12 c). Finally, the dendrogram in figure 4.12 d shows that the biological replicates of hESCs (control) and of DE (treatment) can be clearly separated into distinct groups.

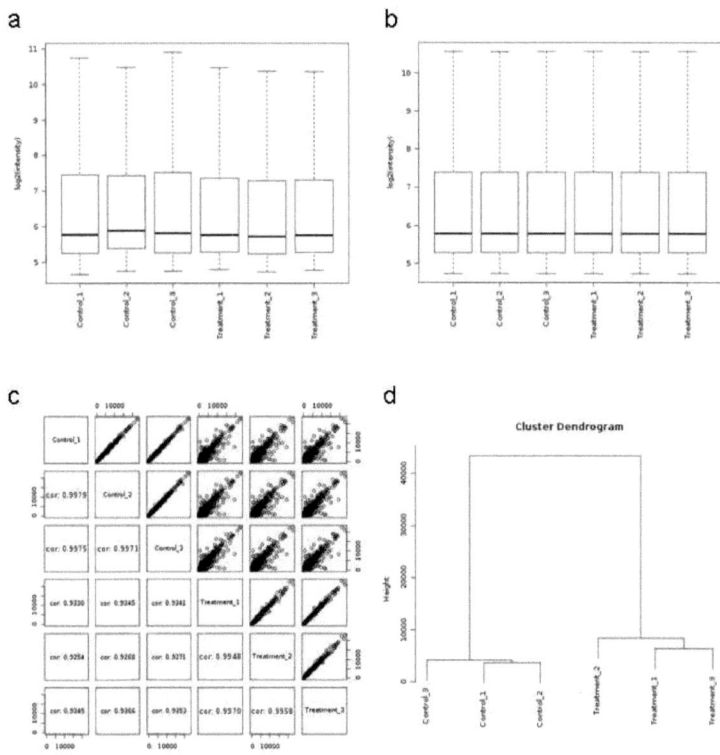

**Figure 4.12: Gene expression quality controls.** The box plots show raw (**a**) and quantile normalized (**b**) Illumina beadarray signal distributions for the three biological replicates of hESCs (control) and DE (treatment). Correlations within groups are always >0.99 and between groups range between 0.92 to 0.93 (**c**). Gene expression profiles of the biological replicates show a good clustering behaviour (**d**).

# 4 Epigenetic dependencies during endodermal differentiation of human ES cells

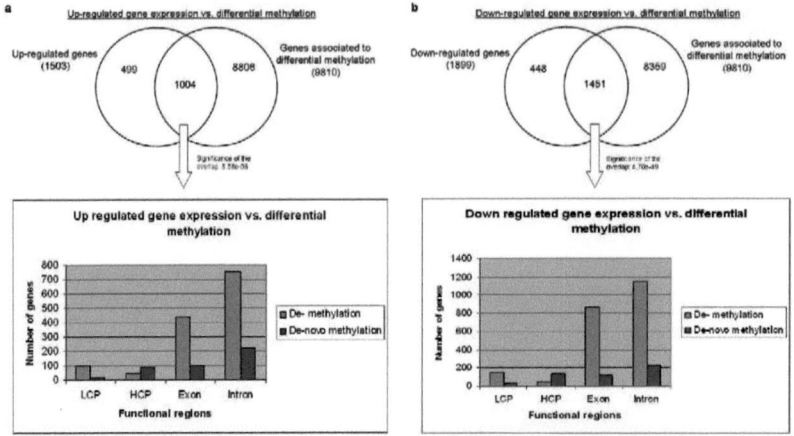

**Figure 4.13: Genetic and epigenetic dependencies.** The figures show the number of up-regulated (a) and down-regulated (b) genes with respect to the number of genes associated to differentially methylated regions. For the genes that are differentially expressed and contain a DMR, the histograms show an overview of the number and location of de- and *de-novo* methylated regions (LCP=low CpG density promoter; HCP=high CpG density promoter).

Differential gene expression was calculated using the limma [105] package, and by setting the level of significance to 0.01. There were 2,129 genes significantly down-regulated and 1,661 genes up-regulated after differentiation.

A gene name-based comparison revealed 15,947 genes common on the utilized Illumina arrays and within the Ensembl [10] gene annotations used for annotating the identified DMRs. Based on these common genes, figure 4.13 shows the overlap between genes that contain at least one identified DMR (de- or *de-novo* methylation) in any of their associated transcript- exons, introns or promoter regions with expression either up-regulated (figure 4.13 a) or down-regulated (figure 4.13 b).

In general, events of differential methylation are significantly associated with up-regulated (p-value=3.58e-06) and down-regulated (p-value=4.78e-49) gene expression patterns. For testing these enrichments, we used the hypergeometric distribution function *phyper* provided within the R framework (www.R-project.org).

However, the histograms in figures 4.13 a and b show similar location distributions over the tested gene-associated functional units of de- and *de-novo* methylation events in both, up- and down regulated genes. Although gene expression changes cannot, in general, be linked to distinct patterns of differential methylation, figure 4.14 shows two examples of *de-novo* methylation events located within the promoter regions of the TFs OCT4/POU5F1 (figure 4.14 a) and STAT5A (figure 4.14 b), which appear along with down-regulation of gene expression in both cases.

## 4.12 Epigenetic effects on the OCT4 regulatory network

In section 4.1, I have presented a core gene regulatory network of OCT4 within the context of maintaining pluripotency in hESCs [17]. I have specifically tested the effect of Activin A treatment on the induction of endodermal differentiation [24] by associating gene expression and DNA methylation for the members of the network (see figure 4.15).

The network illustrates that transcript body associated de- and *de-novo* methylation events occur along with both, up- and down-regulation of gene expression. The results suggest more complex dependencies in the interplay between gene regulation and DNA methylation during endoderm differentiation and of course gastrulation. However, within the context of the network, *de-novo* promoter methylation can be unequivocally associated with OCT4 only and occurs in combination with downregulation of OCT4 expression. Therefore, methylation in the promoter of this core transcription factor may lead to downregulation of OCT4 and subsequently to loss of pluripotency.

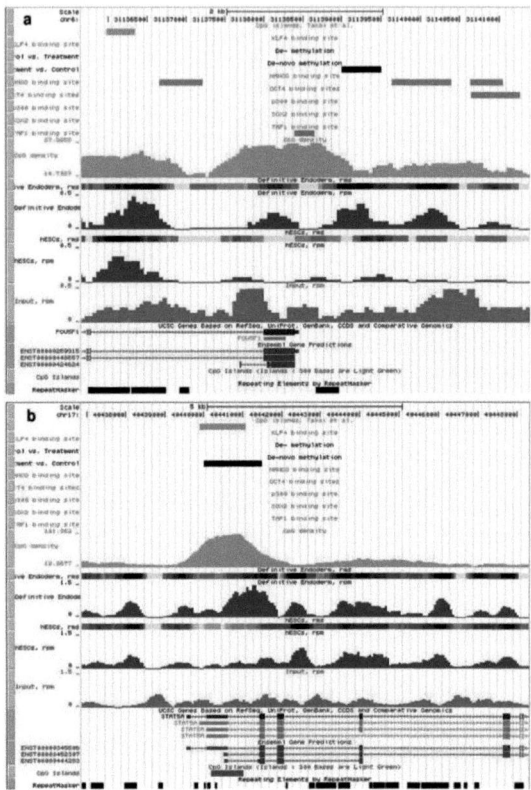

**Figure 4.14: *De-novo* methylation events in the promoter regions of OCT4 and STAT5A.** The figures show example promoter regions of differentially expressed genes visualized by a local copy of the UCSC genome browser [86] (hg19). Included tracks are *rpm* (blue curves) and *rms* (grey blocks) values for hESCs and DE, *rpm* values for input (red curves), de-and *de-novo* methylated regions (black blocks), CpG islands defined by UCSC [86] (dark green blocks at the bottom) and by Takai et al. [97] (light green blocks at the top), CpG densities along the chromosome (green curves, calculated by MEDIPS based on the CpG coupling factors), TFBSs of six TFs [64] (orange blocks; genomic regions were transformed from hg18 to hg19 using UCSCs liftover software [86]), repeat masked regions (black boxes at the bottom), and Ref-Seq [84] and Ensembl [10] transcripts. **a** The promoter region of the down-regulated TF OCT4/POU5F1 including an identified promoter *de-novo* methylation event. **b** The promoter region of the down-regulated TF, STAT5A, including an identified *de-novo* methylation event.

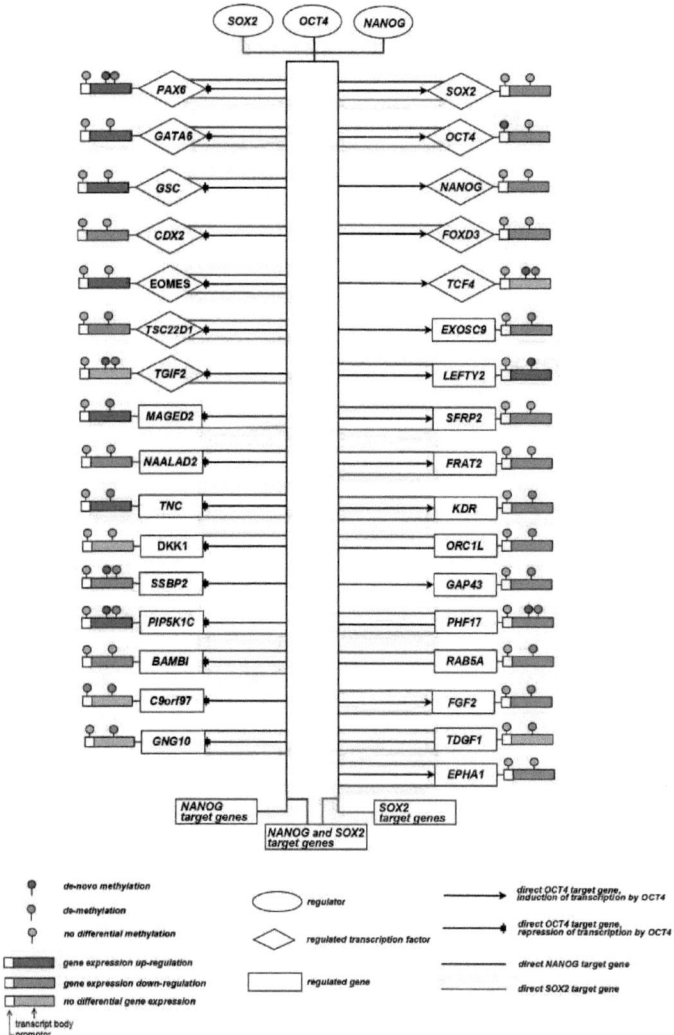

**Figure 4.15: Epigenetic effects on the core OCT4 network.** Core OCT4 network identified within the context of maintaining pluripotency in hESCs [17]. The network illustrates the effect of Activin A treatment as an inducer of endodermal differentiation [24] on gene expression and promoter and transcript body DNA methylation of the individual genes.

# 5 Conclusion

DNA methylation is a crucial epigenetic mechanism involved in normal organismal development and cellular differentiation. Reversible cytosine methylation allows for modifying the DNA without changing the DNA sequence itself and can be passed to successor cells in order to maintain a memory of the cellular state.

Sequencing-based derived DNA methylation data is an emerging source for analyzing epigenetic modifications [54]. The generation of genome-wide methylation data derived from methylated DNA immunoprecipitation followed by sequencing (MeDIP-seq) has become a major tool for epigenetic studies in health and disease. Although MeDIP-seq lacks the ability to discriminate between CpG and non-CpG methylation, it covers nearly as many CpGs per sample genome as the more expensive whole-genome shotgun bisulphite sequencing (WGSBS) approach [54]. An advantage of the MeDIP approach is the generation of unbiased, cost-effective and full-genome methylation levels without the limitations associated with methylation-sensitive restriction enzymes.

However, there was a lack of computational analysis methods of such data, especially in the context of accuracy, sensitivity and speed. As mentioned in section 1.4, it has been shown that MeDIP-derived data needs to be corrected for local CpG densities in order to estimate valid methylation levels [27, 80]. Although there are applicable software available for analyzing MeDIP-Chip data [27, 80], the normalization of MeDIP-seq data is in principle solved [27] but remained disproportional time-consuming. Moreover, there remained several open questions for the analysis of MeDIP-seq data like estimating the number of reads necessary for obtaining a sufficiently covered methylome relative to the size of the genome of interest, the analysis of genome wide covered CpGs, the enrichment of CpG rich short reads relative to the genomic background, as well as the statistical

## 5 Conclusion

identification of differentially methylated regions between different conditions.

Here, an alternative approach for normalizing MeDIP-seq data was presented that is based on the valuable concept of coupling factors presented by Down et al. [27] but out performs computation time by orders of magnitude. In fact, processing of the human chromosome 1 by the BATMAN algorithm [27] takes approximately three days on a modern day one processor server. In contrast, MEDIPS normalizes the full human genome in approximately one hour on the the same hardware. As a proof of principle, we processed the available MeDIP-seq sperm data from Down et al. [27], compared our results to the bisulphite sequencing derived sperm data from the HEP project [30], and show comparable concordance to the results of Down et al [27]. The developed statistical methods are able to cope with the inherent complexity of MeDIP-seq data and out performs computation time of existing methods by orders of magnitude with similar performance. In addition to the normalization approach, novel quality control methods were developed that deal with all of the open questions in the context of MeDIP-seq data as mentioned above.

The developed methods were implemented as an R library (www.R-project.org) named MEDIPS (http://medips.molgen.mpg.de), is suitable for any arbitrary genome available via Bioconductors annotation libraries [33, 77], includes the novel MeDIP-seq specific quality controls, performs data normalization, and enables the identification of differentially methylated regions. MEDIPS offsets the imbalance of sequencing data production and analysis by a simplified and accelerated construction and comparison of full-genome methylation profiles and will hopefully assist further studies aiming to understand and characterize the function of DNA-methylation.

In order to further demonstrate the effectiveness of our analysis tool, we employed human embryonic stem cells (hESCs) as a model for cellular differentiation. Because hESCs and induced pluripotent stem cells (iPSCs) can be induced to differentiate into a wide variety of cell types, these cells hold promise for cell replacement therapy [4]. The transcription factor OCT4 is known as a key regulator for maintaining pluripotency in the mammalian embryo [73, 81]. In order to identify a core OCT4 regulatory network controlling pluripotency in hESCs, I have performed [17] an integrated analysis of heterogeneous high-throughput data along with promoter sequence analysis.

Differentiation of hESCs along the endodermal lineage is induced by treatment with Activin A, a member of the TGF$\beta$ family of ligands [3, 24], resulting in definitive endoderm (DE). Gene expression and DNA-methylation profiles of both cell types were analyzed using the Illumina beadarray platform and MeDIP-seq technologies. Methylomes for hESCs and DE, as well as differentially methylated genomic regions were identified using the presented software. Analogous to Lister et al. [64], we identified a large number of de-methylation events, emphasizing an important role of de-methylation during the differentiation of hESCs. Furthermore, it was shown that in contrast to *de-novo* methylation events, de-methylation is mainly associated with regions of low CpG densities. Finally, I have shown DNA methylation and gene expression alterations in detail for the previously reconstructed core OCT4 network. Here, promoter associated *de-novo* methylation is only observed for OCT4 and occurs in combination with downregulation of OCT4 expression. These observations suggest that OCT4 promoter methylation may lead to downregulation of this core transcription factor and subsequently to loss of pluripotency.

Although in-vivo liver development is specifically characterized by substantial de-methylation, Brunner et al. [15] reported controversial observations on the number of de-methylation and on the enrichment of de- and *de-novo* methylation events at H3K27me3-bound regions and within LCPs when comparing in-vitro and in-vivo hepatic differentiation by a methyl-sensitive restriction enzyme based sequencing approach. However, based on the normalized full-genome MeDIP-seq data of hESCs and DE, consistently with in-vivo hepatic differentiation [15], we observe high numbers of de-methylation events and especially LCPs are specific targets for de-methylation compared to *de-novo* methylation. Although we compared our DMRs to histone modification signals obtained from human T-Cells [7], accordingly to Meissner et al. [72], we observed that H3K4 methylation events (activating mark) are associated with de-methylation events in hESCs.

CpG density and methylation analysis revealed two classes of TFs, namely OCT4, KLF4, and TAF1 on the one hand, and NANOG, SOX2, and p300 on the other hand,

## 5 Conclusion

thus suggesting distinct mechanisms in the interplay between transcription factor binding and DNA methylation. This observation is accompanied by the results obtained when investigating genomic regions identified as TFBSs [64] and as DMRs. Here, TFs whose binding sites show bimodal methylation distributions (these are OCT4, KLF3, and TAF1) overlap more than twice as much with DMRs compared to the TFBSs of the other class of TFs (i.e. NANOG, SOX2, and p300). However, it has to be more extensively analyzed in the future, if these observations indeed reflect different epigenetic regulatory dependencies for different classes of TFs.

As expected, differential methylation is significantly associated with differential gene expression. However, in spite of previous theories [85, 80] suggesting distinct dependencies of methylation changes on gene expression, we did not observe a general trend of gene-associated, region-specific methylation alterations that could explain up- and down-regulation of gene expression. Therefore, we propose that the effect of differential methylation on gene expression has to be further examined with respect to gene-specific locations of putative functional enhancer or silencer regions.

Full genome and base-specific methylomes of hESCs have been previously generated based on >1 billion short reads [64]. In contrast, we have shown that MeDIP-seq coupled with MEDIPS enables the generation of full genome methylation profiles based on approximately 20-30 million uniquely mapped short reads. Although MeDIP-seq data is not base-specific and therefore does not permit distinguishing cytosine methylation in CpG and non-CpG context, we have shown that for regions of interest, the methylation profiles obtained by WGSBS can be correlated to normalized MeDIP-seq data. Nevertheless, Lister and colleagues [64] have shown that in human embryonic stem cells, methylation occurs at cytosines that are not within the CpG context. They have shown that cytosine methylation in none-CpG context accounts for 25% of all methylation events in hESCs. In principle, MEDIPS allows to examine the dependency of the MeDIP-seq signals and any other arbitrary sequence pattern. Therefore, it might be worthwhile to include all cytosines within the reference genome into the normalization process. For this, in addition to CpG specific coupling factors, a coupling vector can be calculated with respect to all cytosines in the reference genome. Testing the effect of MeDIP-seq data normalization based on a coupling vector generated as a weighted combination between e.g. CpG and C coupling factors on the overall data accordance between immunocaptured and bisulphite dependent techniques is only one possible extension that can be immediately approached by the developed MEDIPS software package.

Taken together, in our opinion and in line with D'Amour et al. [24], we propose to further consider in-vitro differentiation of hESCs along the endodermal lineage as a model for endodermal in-vivo development and suggest that MeDIP-seq coupled with MEDIPS is a cost- and time- effective methodology for full genome DNA methylation analysis.

# Bibliography

[1] J. Adjaye, V. Bolton, and M. Monk. Developmental expression of specific genes detected in high-quality cdna libraries from single human preimplantation embryos. *Gene*, 237(2):373–383, Sep 1999.

[2] James Adjaye, John Huntriss, Ralf Herwig, Alia BenKahla, Thore C Brink, Christoph Wierling, Claus Hultschig, Detlef Groth, Marie-Laure Yaspo, Helen M Picton, Roger G Gosden, and Hans Lehrach. Primary differentiation in the human blastocyst: comparative molecular portraits of inner cell mass and trophectoderm cells. *Stem Cells*, 23(10):1514–1525, 2005.

[3] Sadhana Agarwal, Katherine L Holton, and Robert Lanza. Efficient differentiation of functional hepatocytes from human embryonic stem cells. *Stem Cells*, 26(5):1117–1127, May 2008.

[4] Gulsah Altun, Jeanne F Loring, and Louise C Laurent. Dna methylation in embryonic stem cells. *J Cell Biochem*, 109(1):1–6, Jan 2010.

[5] Said Assou, Tanguy Le Carrour, Sylvie Tondeur, Susanne Stroem, Audrey Gabelle, Sophie Marty, Laure Nadal, Veronique Pantesco, Thierry Reme, Jean-Philippe Hugnot, StÃ©phan Gasca, Outi Hovatta, Samir Hamamah, Bernard Klein, and John De Vos. A meta-analysis of human embryonic stem cells transcriptome integrated into a web-based expression atlas. *Stem Cells*, 25(4):961–973, Apr 2007.

[6] Yasmin Babaie, Ralf Herwig, Boris Greber, Thore C Brink, Wasco Wruck, Detlef Groth, Hans Lehrach, Tom Burdon, and James Adjaye. Analysis of oct4-dependent transcriptional networks regulating self-renewal and pluripotency in human embryonic stem cells. *Stem Cells*, 25(2):500–510, Feb 2007.

## Bibliography

[7] Artem Barski, Suresh Cuddapah, Kairong Cui, Tae-Young Roh, Dustin E Schones, Zhibin Wang, Gang Wei, Iouri Chepelev, and Keji Zhao. High-resolution profiling of histone methylations in the human genome. *Cell*, 129(4):823–837, May 2007.

[8] Stephan Beck and Vardhman K Rakyan. The methylome: approaches for global dna methylation profiling. *Trends Genet*, 24(5):231–237, May 2008.

[9] Benjamin P Berman, Daniel J Weisenberger, and Peter W Laird. Locking in on the human methylome. *Nat Biotechnol*, 27(4):341–342, Apr 2009.

[10] Ewan Birney, T. Daniel Andrews, Paul Bevan, Mario Caccamo, Yuan Chen, Laura Clarke, Guy Coates, James Cuff, Val Curwen, Tim Cutts, Thomas Down, Eduardo Eyras, Xose M Fernandez-Suarez, Paul Gane, Brian Gibbins, James Gilbert, Martin Hammond, Hans-Rudolf Hotz, Vivek Iyer, Kerstin Jekosch, Andreas Kahari, Arek Kasprzyk, Damian Keefe, Stephen Keenan, Heikki Lehvaslaiho, Graham McVicker, Craig Melsopp, Patrick Meidl, Emmanuel Mongin, Roger Pettett, Simon Potter, Glenn Proctor, Mark Rae, Steve Searle, Guy Slater, Damian Smedley, James Smith, Will Spooner, Arne Stabenau, James Stalker, Roy Storey, Abel Ureta-Vidal, K. Cara Woodwark, Graham Cameron, Richard Durbin, Anthony Cox, Tim Hubbard, and Michele Clamp. An overview of ensembl. *Genome Res*, 14(5):925–928, May 2004.

[11] Christoph Bock, Konstantin Halachev, Joachim BÃŒch, and Thomas Lengauer. Epigraph: user-friendly software for statistical analysis and prediction of (epi)genomic data. *Genome Biol*, 10(2):R14, 2009.

[12] V. Botquin, H. Hess, G. Fuhrmann, C. Anastassiadis, M. K. Gross, G. Vriend, and H. R. Schoeler. New pou dimer configuration mediates antagonistic control of an osteopontin preimplantation enhancer by oct-4 and sox-2. *Genes Dev*, 12(13):2073–2090, Jul 1998.

[13] Laurie A Boyer, Tong Ihn Lee, Megan F Cole, Sarah E Johnstone, Stuart S Levine, Jacob P Zucker, Matthew G Guenther, Roshan M Kumar, Heather L Murray, Richard G Jenner, David K Gifford, Douglas A Melton, Rudolf Jaenisch, and Richard A Young. Core transcriptional regulatory circuitry in human embryonic stem cells. *Cell*, 122(6):947–956, Sep 2005.

## Bibliography

[14] Alan P Boyle, Justin Guinney, Gregory E Crawford, and Terrence S Furey. F-seq: a feature density estimator for high-throughput sequence tags. *Bioinformatics*, 24(21):2537–2538, Nov 2008.

[15] Alayne L Brunner, David S Johnson, Si Wan Kim, Anton Valouev, Timothy E Reddy, Norma F Neff, Elizabeth Anton, Catherine Medina, Loan Nguyen, Eric Chiao, Chuba B Oyolu, Gary P Schroth, Devin M Absher, Julie C Baker, and Richard M Myers. Distinct dna methylation patterns characterize differentiated human embryonic stem cells and developing human fetal liver. *Genome Res*, 19(6):1044–1056, Jun 2009.

[16] Elayne M Chan, Sutheera Ratanasirintrawoot, In-Hyun Park, Philip D Manos, Yuin-Han Loh, Hongguang Huo, Justine D Miller, Odelya Hartung, Junsung Rho, Tan A Ince, George Q Daley, and Thorsten M Schlaeger. Live cell imaging distinguishes bona fide human ips cells from partially reprogrammed cells. *Nat Biotechnol*, 27(11):1033–1037, Nov 2009.

[17] Lukas Chavez, Abha S Bais, Martin Vingron, Hans Lehrach, James Adjaye, and Ralf Herwig. In silico identification of a core regulatory network of oct4 in human embryonic stem cells using an integrated approach. *BMC Genomics*, 10:314, 2009.

[18] Joon-Lin Chew, Yuin-Han Loh, Wensheng Zhang, Xi Chen, Wai-Leong Tam, Leng-Siew Yeap, Pin Li, Yen-Sin Ang, Bing Lim, Paul Robson, and Huck-Hui Ng. Reciprocal transcriptional regulation of pou5f1 and sox2 via the oct4/sox2 complex in embryonic stem cells. *Mol Cell Biol*, 25(14):6031–6046, Jul 2005.

[19] George M Church. Genomes for all. *Sci Am*, 294(1):46–54, Jan 2006.

[20] S. J. Clark, J. Harrison, C. L. Paul, and M. Frommer. High sensitivity mapping of methylated cytosines. *Nucleic Acids Res*, 22(15):2990–2997, Aug 1994.

[21] J. M. Claverie. Computational methods for the identification of differential and coordinated gene expression. *Hum Mol Genet*, 8(10):1821–1832, 1999.

[22] Shawn J Cokus, Suhua Feng, Xiaoyu Zhang, Zugen Chen, Barry Merriman, Christian D Haudenschild, Sriharsa Pradhan, Stanley F Nelson, Matteo Pellegrini, and

## Bibliography

Steven E Jacobsen. Shotgun bisulphite sequencing of the arabidopsis genome reveals dna methylation patterning. *Nature*, 452(7184):215–219, Mar 2008.

[23] Philippe Collas. The current state of chromatin immunoprecipitation. *Mol Biotechnol*, Jan 2010.

[24] Kevin A D'Amour, Alan D Agulnick, Susan Eliazer, Olivia G Kelly, Evert Kroon, and Emmanuel E Baetge. Efficient differentiation of human embryonic stem cells to definitive endoderm. *Nat Biotechnol*, 23(12):1534–1541, Dec 2005.

[25] Michael P Dempsey, Joseph Nietfeldt, Jacques Ravel, Steven Hinrichs, Robert Crawford, and Andrew K Benson. Paired-end sequence mapping detects extensive genomic rearrangement and translocation during divergence of francisella tularensis subsp. tularensis and francisella tularensis subsp. holarctica populations. *J Bacteriol*, 188(16):5904–5914, Aug 2006.

[26] Jie Deng, Robert Shoemaker, Bin Xie, Athurva Gore, Emily M LeProust, Jessica Antosiewicz-Bourget, Dieter Egli, Nimet Maherali, In-Hyun Park, Junying Yu, George Q Daley, Kevin Eggan, Konrad Hochedlinger, James Thomson, Wei Wang, Yuan Gao, and Kun Zhang. Targeted bisulfite sequencing reveals changes in dna methylation associated with nuclear reprogramming. *Nat Biotechnol*, 27(4):353–360, Apr 2009.

[27] Thomas A Down, Vardhman K Rakyan, Daniel J Turner, Paul Flicek, Heng Li, Eugene Kulesha, Stefan Graef, Nathan Johnson, Javier Herrero, Eleni M Tomazou, Natalie P Thorne, Liselotte Baeckdahl, Marlis Herberth, Kevin L Howe, David K Jackson, Marcos M Miretti, John C Marioni, Ewan Birney, Tim J P Hubbard, Richard Durbin, Simon Tavare, and Stephan Beck. A bayesian deconvolution strategy for immunoprecipitation-based dna methylome analysis. *Nat Biotechnol*, 26(7):779–785, Jul 2008.

[28] John J Dunn, Sean R McCorkle, Logan Everett, and Carl W Anderson. Paired-end genomic signature tags: a method for the functional analysis of genomes and epigenomes. *Genet Eng (N Y)*, 28:159–173, 2007.

[29] Mark J Dunning, Mike L Smith, Matthew E Ritchie, and Simon Tavaré.

*Bibliography*

beadarray: R classes and methods for illumina bead-based data. *Bioinformatics*, 23(16):2183–2184, Aug 2007.

[30] Florian Eckhardt, Joern Lewin, Rene Cortese, Vardhman K Rakyan, John Attwood, Matthias Burger, John Burton, Tony V Cox, Rob Davies, Thomas A Down, Carolina Haefliger, Roger Horton, Kevin Howe, David K Jackson, Jan Kunde, Christoph Koenig, Jennifer Liddle, David Niblett, Thomas Otto, Roger Pettett, Stefanie Seemann, Christian Thompson, Tony West, Jane Rogers, Alex Olek, Kurt Berlin, and Stephan Beck. Dna methylation profiling of human chromosomes 6, 20 and 22. *Nat Genet*, 38(12):1378–1385, Dec 2006.

[31] M. Frommer, L. E. McDonald, D. S. Millar, C. M. Collis, F. Watt, G. W. Grigg, P. L. Molloy, and C. L. Paul. A genomic sequencing protocol that yields a positive display of 5-methylcytosine residues in individual dna strands. *Proc Natl Acad Sci U S A*, 89(5):1827–1831, Mar 1992.

[32] M. Gardiner-Garden and M. Frommer. Cpg islands in vertebrate genomes. *J Mol Biol*, 196(2):261–282, Jul 1987.

[33] Robert C Gentleman, Vincent J Carey, Douglas M Bates, Ben Bolstad, Marcel Dettling, Sandrine Dudoit, Byron Ellis, Laurent Gautier, Yongchao Ge, Jeff Gentry, Kurt Hornik, Torsten Hothorn, Wolfgang Huber, Stefano Iacus, Rafael Irizarry, Friedrich Leisch, Cheng Li, Martin Maechler, Anthony J Rossini, Gunther Sawitzki, Colin Smith, Gordon Smyth, Luke Tierney, Jean Y H Yang, and Jianhua Zhang. Bioconductor: open software development for computational biology and bioinformatics. *Genome Biol*, 5(10):R80, 2004.

[34] Boris Greber, Hans Lehrach, and James Adjaye. Fibroblast growth factor 2 modulates transforming growth factor beta signaling in mouse embryonic fibroblasts and human escs (hescs) to support hesc self-renewal. *Stem Cells*, 25(2):455–464, Feb 2007.

[35] Neil Hall. Advanced sequencing technologies and their wider impact in microbiology. *J Exp Biol*, 210(Pt 9):1518–1525, May 2007.

[36] Izuho Hatada. Emerging technologies for genome-wide dna methylation profiling in cancer. *Crit Rev Oncog*, 12(3-4):205–223, Dec 2006.

## Bibliography

[37] Hikoya Hayatsu. Discovery of bisulfite-mediated cytosine conversion to uracil, the key reaction for dna methylation analysis–a personal account. *Proc Jpn Acad Ser B Phys Biol Sci*, 84(8):321–330, 2008.

[38] R. Herwig, P. Aanstad, M. Clark, and H. Lehrach. Statistical evaluation of differential expression on cdna nylon arrays with replicated experiments. *Nucleic Acids Res*, 29(23):E117, Dec 2001.

[39] Danwei Huangfu, Kenji Osafune, Rene Maehr, Wenjun Guo, Astrid Eijkelenboom, Shuibing Chen, Whitney Muhlestein, and Douglas A Melton. Induction of pluripotent stem cells from primary human fibroblasts with only oct4 and sox2. *Nat Biotechnol*, 26(11):1269–1275, Nov 2008.

[40] T. Ideker, V. Thorsson, A. F. Siegel, and L. E. Hood. Testing for differentially-expressed genes by maximum-likelihood analysis of microarray data. *J Comput Biol*, 7(6):805–817, 2000.

[41] Rafael A Irizarry, Christine Ladd-Acosta, Bo Wen, Zhijin Wu, Carolina Montano, Patrick Onyango, Hengmi Cui, Kevin Gabo, Michael Rongione, Maree Webster, Hong Ji, James B Potash, Sarven Sabunciyan, and Andrew P Feinberg. The human colon cancer methylome shows similar hypo- and hypermethylation at conserved tissue-specific cpg island shores. *Nat Genet*, 41(2):178–186, Feb 2009.

[42] Filipe V Jacinto, Esteban Ballestar, and Manel Esteller. Methyl-dna immunoprecipitation (medip): hunting down the dna methylome. *Biotechniques*, 44(1):35, 37, 39 passim, Jan 2008.

[43] Rudolf Jaenisch and Adrian Bird. Epigenetic regulation of gene expression: how the genome integrates intrinsic and environmental signals. *Nat Genet*, 33 Suppl:245–254, Mar 2003.

[44] Jeffrey A Jeddeloh, John M Greally, and Oliver J Rando. Reduced-representation methylation mapping. *Genome Biol*, 9(8):231, 2008.

[45] Hongkai Ji, Hui Jiang, Wenxiu Ma, David S Johnson, Richard M Myers, and Wing H Wong. An integrated software system for analyzing chip-chip and chip-seq data. *Nat Biotechnol*, 26(11):1293–1300, Nov 2008.

## Bibliography

[46] W. Evan Johnson, Wei Li, Clifford A Meyer, Raphael Gottardo, Jason S Carroll, Myles Brown, and X. Shirley Liu. Model-based analysis of tiling-arrays for chip-chip. *Proc Natl Acad Sci U S A*, 103(33):12457–12462, Aug 2006.

[47] Peter A Jones and Stephen B Baylin. The epigenomics of cancer. *Cell*, 128(4):683–692, Feb 2007.

[48] Justyna Jozefczuk, Harald Stachelscheid, Lukas Chavez, Ralf Herwig, Hans Lehrach, Katrin Zeilinger, Joerg C Gerlach, and James Adjaye. Molecular characterization of cultured adult human liver progenitor cells. *Tissue Eng Part C Methods*, Jan 2010.

[49] Marc Jung, Hedi Peterson, Lukas Chavez, Pascal Kahlem, Hans Lehrach, Jaak Vilo, and James Adjaye. A data integration approach to mapping oct4 gene regulatory networks operative in embryonic stem cells and embryonal carcinoma cells. *PLoS One*, 5(5):e10709, 2010.

[50] Irina Klimanskaya, Young Chung, Sandy Becker, Shi-Jiang Lu, and Robert Lanza. Human embryonic stem cell lines derived from single blastomeres. *Nature*, 444(7118):481–485, Nov 2006.

[51] Jan O Korbel, Alexander Eckehart Urban, Jason P Affourtit, Brian Godwin, Fabian Grubert, Jan Fredrik Simons, Philip M Kim, Dean Palejev, Nicholas J Carriero, Lei Du, Bruce E Taillon, Zhoutao Chen, Andrea Tanzer, A. C Eugenia Saunders, Jianxiang Chi, Fengtang Yang, Nigel P Carter, Matthew E Hurles, Sherman M Weissman, Timothy T Harkins, Mark B Gerstein, Michael Egholm, and Michael Snyder. Paired-end mapping reveals extensive structural variation in the human genome. *Science*, 318(5849):420–426, Oct 2007.

[52] Kenneth Kuhn, Shawn C Baker, Eugene Chudin, Minh-Ha Lieu, Steffen Oeser, Holly Bennett, Philippe Rigault, David Barker, Timothy K McDaniel, and Mark S Chee. A novel, high-performance random array platform for quantitative gene expression profiling. *Genome Res*, 14(11):2347–2356, Nov 2004.

[53] Peter W Laird. The power and the promise of dna methylation markers. *Nat Rev Cancer*, 3(4):253–266, Apr 2003.

## Bibliography

[54] Peter W Laird. Principles and challenges of genome-wide dna methylation analysis. *Nat Rev Genet*, 11(3):191–203, Feb 2010.

[55] Ben Langmead, Cole Trapnell, Mihai Pop, and Steven L Salzberg. Ultrafast and memory-efficient alignment of short dna sequences to the human genome. *Genome Biol*, 10(3):R25, 2009.

[56] Benjamin P Lewis, Christopher B Burge, and David P Bartel. Conserved seed pairing, often flanked by adenosines, indicates that thousands of human genes are microrna targets. *Cell*, 120(1):15–20, Jan 2005.

[57] Heng Li, Jue Ruan, and Richard Durbin. Mapping short dna sequencing reads and calling variants using mapping quality scores. *Genome Res*, 18(11):1851–1858, Nov 2008.

[58] Ning Li, Mingzhi Ye, Yingrui Li, Zhixiang Yan, Lee M Butcher, Jihua Sun, Xu Han, Quan Chen, Xiuqing Zhang, and Jun Wang. Whole genome dna methylation analysis based on high throughput sequencing technology. *Methods*, Apr 2010.

[59] Wei Li, Clifford A Meyer, and X. Shirley Liu. A hidden markov model for analyzing chip-chip experiments on genome tiling arrays and its application to p53 binding sequences. *Bioinformatics*, 21 Suppl 1:i274–i282, Jun 2005.

[60] J. D. Lieb, S. Beck, M. L. Bulyk, P. Farnham, N. Hattori, S. Henikoff, X. S. Liu, K. Okumura, K. Shiota, T. Ushijima, and J. M. Greally. Applying whole-genome studies of epigenetic regulation to study human disease. *Cytogenet Genome Res*, 114(1):1–15, 2006.

[61] Simon M Lin, Pan Du, Wolfgang Huber, and Warren A Kibbe. Model-based variance-stabilizing transformation for illumina microarray data. *Nucleic Acids Res*, 36(2):e11, Feb 2008.

[62] Ryan Lister and Joseph R Ecker. Finding the fifth base: genome-wide sequencing of cytosine methylation. *Genome Res*, 19(6):959–966, Jun 2009.

[63] Ryan Lister, Ronan C O'Malley, Julian Tonti-Filippini, Brian D Gregory, Charles C Berry, A. Harvey Millar, and Joseph R Ecker. Highly integrated single-base resolution maps of the epigenome in arabidopsis. *Cell*, 133(3):523–536, May 2008.

## Bibliography

[64] Ryan Lister, Mattia Pelizzola, Robert H Dowen, R. David Hawkins, Gary Hon, Julian Tonti-Filippini, Joseph R Nery, Leonard Lee, Zhen Ye, Que-Minh Ngo, Lee Edsall, Jessica Antosiewicz-Bourget, Ron Stewart, Victor Ruotti, A. Harvey Millar, James A Thomson, Bing Ren, and Joseph R Ecker. Human dna methylomes at base resolution show widespread epigenomic differences. *Nature*, 462(7271):315–322, Nov 2009.

[65] D. J. Lockhart, H. Dong, M. C. Byrne, M. T. Follettie, M. V. Gallo, M. S. Chee, M. Mittmann, C. Wang, M. Kobayashi, H. Horton, and E. L. Brown. Expression monitoring by hybridization to high-density oligonucleotide arrays. *Nat Biotechnol*, 14(13):1675–1680, Dec 1996.

[66] D. J. Lockhart and E. A. Winzeler. Genomics, gene expression and dna arrays. *Nature*, 405(6788):827–836, Jun 2000.

[67] Desmond S Lun, Ashley Sherrid, Brian Weiner, David R Sherman, and James E Galagan. A blind deconvolution approach to high-resolution mapping of transcription factor binding sites from chip-seq data. *Genome Biol*, 10(12):R142, 2009.

[68] V. Matys, E. Fricke, R. Geffers, E. Goessling, M. Haubrock, R. Hehl, K. Hornischer, D. Karas, A. E. Kel, O. V. Kel-Margoulis, D-U. Kloos, S. Land, B. Lewicki-Potapov, H. Michael, R. MÃŒnch, I. Reuter, S. Rotert, H. Saxel, M. Scheer, S. Thiele, and E. Wingender. Transfac: transcriptional regulation, from patterns to profiles. *Nucleic Acids Res*, 31(1):374–378, Jan 2003.

[69] A. M. Maxam and W. Gilbert. A new method for sequencing dna. *Proc Natl Acad Sci U S A*, 74(2):560–564, Feb 1977.

[70] R. R. Meehan, J. D. Lewis, and A. P. Bird. Characterization of mecp2, a vertebrate dna binding protein with affinity for methylated dna. *Nucleic Acids Res*, 20(19):5085–5092, Oct 1992.

[71] Alexander Meissner, Andreas Gnirke, George W Bell, Bernard Ramsahoye, Eric S Lander, and Rudolf Jaenisch. Reduced representation bisulfite sequencing for comparative high-resolution dna methylation analysis. *Nucleic Acids Res*, 33(18):5868–5877, 2005.

*Bibliography*

[72] Alexander Meissner, Tarjei S Mikkelsen, Hongcang Gu, Marius Wernig, Jacob Hanna, Andrey Sivachenko, Xiaolan Zhang, Bradley E Bernstein, Chad Nusbaum, David B Jaffe, Andreas Gnirke, Rudolf Jaenisch, and Eric S Lander. Genome-scale dna methylation maps of pluripotent and differentiated cells. *Nature*, 454(7205):766–770, Aug 2008.

[73] J. Nichols, B. Zevnik, K. Anastassiadis, H. Niwa, D. Klewe-Nebenius, I. Chambers, H. Schoeler, and A. Smith. Formation of pluripotent stem cells in the mammalian embryo depends on the pou transcription factor oct4. *Cell*, 95(3):379–391, Oct 1998.

[74] M. Nishimoto, A. Fukushima, A. Okuda, and M. Muramatsu. The gene for the embryonic stem cell coactivator utf1 carries a regulatory element which selectively interacts with a complex composed of oct-3/4 and sox-2. *Mol Cell Biol*, 19(8):5453–5465, Aug 1999.

[75] Hitoshi Niwa, Yayoi Toyooka, Daisuke Shimosato, Dan Strumpf, Kadue Takahashi, Rika Yagi, and Janet Rossant. Interaction between oct3/4 and cdx2 determines trophectoderm differentiation. *Cell*, 123(5):917–929, Dec 2005.

[76] V. Orlando. Mapping chromosomal proteins in vivo by formaldehyde-crosslinked-chromatin immunoprecipitation. *Trends Biochem Sci*, 25(3):99–104, Mar 2000.

[77] Herve Pages. *BSgenome: Infrastructure for Biostrings-based genome data packages*. R package version 1.14.2.

[78] Dmitri Parkhomchuk, Tatiana Borodina, Vyacheslav Amstislavskiy, Maria Banaru, Linda Hallen, Sylvia Krobitsch, Hans Lehrach, and Alexey Soldatov. Transcriptome analysis by strand-specific sequencing of complementary dna. *Nucleic Acids Res*, 37(18):e123, Oct 2009.

[79] C. L. Paul and S. J. Clark. Cytosine methylation: quantitation by automated genomic sequencing and genescan analysis. *Biotechniques*, 21(1):126–133, Jul 1996.

[80] Mattia Pelizzola, Yasuo Koga, Alexander Eckehart Urban, Michael Krauthammer, Sherman Weissman, Ruth Halaban, and Annette M Molinaro. Medme: an experimental and analytical methodology for the estimation of dna methylation levels

based on microarray derived medip-enrichment. *Genome Res*, 18(10):1652–1659, Oct 2008.

[81] M. Pesce, M. K. Gross, and H. R. Schoeler. In line with our ancestors: Oct-4 and the mammalian germ. *Bioessays*, 20(9):722–732, Sep 1998.

[82] M. Pesce and H. R. Schoeler. Oct-4: gatekeeper in the beginnings of mammalian development. *Stem Cells*, 19(4):271–278, 2001.

[83] Kyle R Pomraning, Kristina M Smith, and Michael Freitag. Genome-wide high throughput analysis of dna methylation in eukaryotes. *Methods*, 47(3):142–150, Mar 2009.

[84] Kim D Pruitt, Tatiana Tatusova, and Donna R Maglott. Ncbi reference sequences (refseq): a curated non-redundant sequence database of genomes, transcripts and proteins. *Nucleic Acids Res*, 35(Database issue):D61–D65, Jan 2007.

[85] Vardhman K Rakyan, Thomas A Down, Natalie P Thorne, Paul Flicek, Eugene Kulesha, Stefan Graef, Eleni M Tomazou, Liselotte Baeckdahl, Nathan Johnson, Marlis Herberth, Kevin L Howe, David K Jackson, Marcos M Miretti, Heike Fiegler, John C Marioni, Ewan Birney, Tim J P Hubbard, Nigel P Carter, Simon Tavare, and Stephan Beck. An integrated resource for genome-wide identification and analysis of human tissue-specific differentially methylated regions (tdmrs). *Genome Res*, 18(9):1518–1529, Sep 2008.

[86] Brooke Rhead, Donna Karolchik, Robert M Kuhn, Angie S Hinrichs, Ann S Zweig, Pauline A Fujita, Mark Diekhans, Kayla E Smith, Kate R Rosenbloom, Brian J Raney, Andy Pohl, Michael Pheasant, Laurence R Meyer, Katrina Learned, Fan Hsu, Jennifer Hillman-Jackson, Rachel A Harte, Belinda Giardine, Timothy R Dreszer, Hiram Clawson, Galt P Barber, David Haussler, and W. James Kent. The ucsc genome browser database: update 2010. *Nucleic Acids Res*, 38(Database issue):D613–D619, Jan 2010.

[87] David J Rodda, Joon-Lin Chew, Leng-Hiong Lim, Yuin-Han Loh, Bei Wang, Huck-Hui Ng, and Paul Robson. Transcriptional regulation of nanog by oct4 and sox2. *J Biol Chem*, 280(26):24731–24737, Jul 2005.

## Bibliography

[88] Joel Rozowsky, Ghia Euskirchen, Raymond K Auerbach, Zhengdong D Zhang, Theodore Gibson, Robert Bjornson, Nicholas Carriero, Michael Snyder, and Mark B Gerstein. Peakseq enables systematic scoring of chip-seq experiments relative to controls. *Nat Biotechnol*, 27(1):66–75, Jan 2009.

[89] Thomas Sandmann, Janus S Jakobsen, and Eileen E M Furlong. Chip-on-chip protocol for genome-wide analysis of transcription factor binding in drosophila melanogaster embryos. *Nat Protoc*, 1(6):2839–2855, 2006.

[90] F. Sanger and A. R. Coulson. A rapid method for determining sequences in dna by primed synthesis with dna polymerase. *J Mol Biol*, 94(3):441–448, May 1975.

[91] F. Sanger, S. Nicklen, and A. R. Coulson. Dna sequencing with chain-terminating inhibitors. *Proc Natl Acad Sci U S A*, 74(12):5463–5467, Dec 1977.

[92] M. Schena, D. Shalon, R. W. Davis, and P. O. Brown. Quantitative monitoring of gene expression patterns with a complementary dna microarray. *Science*, 270(5235):467–470, Oct 1995.

[93] H. R. Schoeler, R. Balling, A. K. Hatzopoulos, N. Suzuki, and P. Gruss. Octamer binding proteins confer transcriptional activity in early mouse embryogenesis. *EMBO J*, 8(9):2551–2557, Sep 1989.

[94] Stephan C Schuster. Next-generation sequencing transforms today's biology. *Nat Methods*, 5(1):16–18, Jan 2008.

[95] Kazutoshi Takahashi, Koji Tanabe, Mari Ohnuki, Megumi Narita, Tomoko Ichisaka, Kiichiro Tomoda, and Shinya Yamanaka. Induction of pluripotent stem cells from adult human fibroblasts by defined factors. *Cell*, 131(5):861–872, Nov 2007.

[96] Kazutoshi Takahashi and Shinya Yamanaka. Induction of pluripotent stem cells from mouse embryonic and adult fibroblast cultures by defined factors. *Cell*, 126(4):663–676, Aug 2006.

[97] Daiya Takai and Peter A Jones. Comprehensive analysis of cpg islands in human chromosomes 21 and 22. *Proc Natl Acad Sci U S A*, 99(6):3740–3745, Mar 2002.

## Bibliography

[98] J. A. Thomson, J. Itskovitz-Eldor, S. S. Shapiro, M. A. Waknitz, J. J. Swiergiel, V. S. Marshall, and J. M. Jones. Embryonic stem cell lines derived from human blastocysts. *Science*, 282(5391):1145–1147, Nov 1998.

[99] Joern Toedling, Oleg Skylar, Oleg Sklyar, Tammo Krueger, Jenny J Fischer, Silke Sperling, and Wolfgang Huber. Ringo–an r/bioconductor package for analyzing chip-chip readouts. *BMC Bioinformatics*, 8:221, 2007.

[100] Toshikazu Ushijima. Detection and interpretation of altered methylation patterns in cancer cells. *Nat Rev Cancer*, 5(3):223–231, Mar 2005.

[101] Anton Valouev, David S Johnson, Andreas Sundquist, Catherine Medina, Elizabeth Anton, Serafim Batzoglou, Richard M Myers, and Arend Sidow. Genome-wide analysis of transcription factor binding sites based on chip-seq data. *Nat Methods*, 5(9):829–834, Sep 2008.

[102] R. Y. Wang, C. W. Gehrke, and M. Ehrlich. Comparison of bisulfite modification of 5-methyldeoxycytidine and deoxycytidine residues. *Nucleic Acids Res*, 8(20):4777–4790, Oct 1980.

[103] Michael Weber, Jonathan J Davies, David Wittig, Edward J Oakeley, Michael Haase, Wan L Lam, and Dirk Schuebeler. Chromosome-wide and promoter-specific analyses identify sites of differential dna methylation in normal and transformed human cells. *Nat Genet*, 37(8):853–862, Aug 2005.

[104] Marius Wernig, Alexander Meissner, Ruth Foreman, Tobias Brambrink, Manching Ku, Konrad Hochedlinger, Bradley E Bernstein, and Rudolf Jaenisch. In vitro reprogramming of fibroblasts into a pluripotent es-cell-like state. *Nature*, 448(7151):318–324, Jul 2007.

[105] James M Wettenhall and Gordon K Smyth. limmagui: a graphical user interface for linear modeling of microarray data. *Bioinformatics*, 20(18):3705–3706, Dec 2004.

[106] Ian M Wilson, Jonathan J Davies, Michael Weber, Carolyn J Brown, Carlos E Alvarez, Calum MacAulay, Dirk SchÃŒbeler, and Wan L Lam. Epigenomics: mapping the methylome. *Cell Cycle*, 5(2):155–158, Jan 2006.

[107] Junying Yu, Maxim A Vodyanik, Kim Smuga-Otto, Jessica Antosiewicz-Bourget, Jennifer L Frane, Shulan Tian, Jeff Nie, Gudrun A Jonsdottir, Victor Ruotti, Ron Stewart, Igor I Slukvin, and James A Thomson. Induced pluripotent stem cell lines derived from human somatic cells. *Science*, 318(5858):1917–1920, Dec 2007.

[108] Xiaoyu Zhang, Junshi Yazaki, Ambika Sundaresan, Shawn Cokus, Simon W-L Chan, Huaming Chen, Ian R Henderson, Paul Shinn, Matteo Pellegrini, Steve E Jacobsen, and Joseph R Ecker. Genome-wide high-resolution mapping and functional analysis of dna methylation in arabidopsis. *Cell*, 126(6):1189–1201, Sep 2006.

# Notation and Abbreviations

| | |
|---|---|
| hESCs | Human Embryonic Stem Cells |
| iPSCs | Induced Pluripotent Stem Cells |
| DE | Definitive Endoderm |
| Input | Control Sample consisting of sheared but not immunoprecipitated DNA |
| TF | Transcription Factor |
| TFBS | Transcription Factor Binding Site |
| DMR | Differentially Methylated Region |
| ROI | Region of Interest |
| MeDIP | Methylated DNA Immunoprecipitation |
| ChIP | Chromatin Immunoprecipitation |
| CpG | Neighboring Cytosine (C) and Guanine (G) located on a DNA strand |
| mCpG | Methylated CpG Di-nucleotide |
| 5mC | Methylated Cytosine |
| CpGw | Weighted Sum of CpGs |
| mCpGw | Weighted Sum of Methylated CpGs |
| HCP | High CpG Density Promoter |
| LCP | Low CpG Density Promoter |
| WGSBS | Whole Genome Shotgun Bisulphite Sequencing |
| RRBS | Reduced Representation Bisulphite Sequencing |
| nt | Nucleotide |
| TSS | Transcription Start Site |
| CF | Coupling Factor |
| OLS | Ordinary Least Square Estimate |
| RPM | Reads Per Million |
| RMS | Relative Methylation Score |
| AMS | Absolute Methylation Score |

# Publications

Jozefczuk J, Prigione A, Chavez L, Adjaye J. *Comparative analysis of human embryonic stem cell and induced pluripotent stem cell-derived hepatocyte-like cells reveals current drawbacks and possible strategies for improved differentiation*, **Stem Cells Dev.** 2011 Jul;20(7):1259-75. Epub 2011 Jan 24

Lukas Chavez, Justyna Jozefczuk, Christina Grimm, Jörn Dietrich, Bernd Timmermann, Ralf Herwig, James Adjaye. *Computational analysis of genome-wide DNA-methylation during the differentiation of human embryonic stem cells along the endodermal lineage*, **Genome Research** 2010 Oct;20(10):1441-50. Epub 2010 Aug 27.

Jung M, Peterson H, Chavez L, Pascal Kahlem, Hans Lehrach, Jaak Vilo, and James Adjaye. *A data integration approach to mapping oct4 gene regulatory networks operative in embryonic stem cells and embryonal carcinoma cells.* **PLoS One**, 5(5):e10709, 2010.

Jozefczuk J*, Stachelscheid H*, Chavez L*, Herwig R, Lehrach H, Zeilinger K, Gerlach JC, Adjaye J. *Molecular Characterization of Cultured Adult Human Liver Progenitor Cells.* **Tissue Eng Part C Methods**, 2010 Jan 11. (* equally contributed)

Lukas Chavez, Abha S Bais, Martin Vingron, Hans Lehrach, James Adjaye, and Ralf Herwig. *In silico identification of a core regulatory network of OCT4 in human embryonic stem cells using an integrated approach.* **BMC Genomics**, 10:314, 2009. (highly accessed)

Ralf Herwig, Johannes Schuchhardt, Lukas Chavez and Hans Lehrach. *Analyse von Biochips: Von der Sequenz zum System*, in Grundlagen der Molekularen Medizin, **Springer**

*Publications*

**Verlag**, Berlin (2007)

# Acknowledgments

This work was carried out at the *Bioinformatics* group of the Department of Vertebrate Genomics at the Max-Planck-Institute for Molecular Genetics in Berlin. I thank all past end present colleagues for the good working atmosphere and fruitful discussions.

Especially, I am grateful to my supervisor *Ralf Herwig* for giving me the opportunity to work in and learn from the group he established. The freedom of research and available guidance and support whenever necessary was an exceptional environment for independent but well supervised scientific work. I thank *James Adjaye* for introducing me to stem cell research. His experiences and advises guided me to the most up-to-date topics in the fields of pluripotency and reprogramming. The profound methodological experiences in the field of computational biology available in the Herwig lab together with the experimental capabilities of the Adjaye lab were an excellent working environment at the interface of data generation, analysis, and interpretation. In addition, I am thankful for the comprehensive working environments of the departments Vertebrate Genomics and Computational Molecular Biology at the Max Planck Institute for Molecular Genetics established by Hans Lehrach and Martin Vingron.

Because investigation of methylation alterations during differentiation of human embryonic stem cells would not be possible without according wet-lab experiments, special thanks go to Justyna Jozefczuk, Christina Grimm, Michal Schweiger, Bernd Timmerman, and Jörn Walter for derivation of definitive endoderm from stem cells, MeDIP experiments, MeDIP and bisulphite sequencing, and for extensive discussions on their scientific fields. I thank Jörn Dietrich for contributing assistance on the development of the MEDIPS software package. For proofreading of this book and of the MEDIPS manual, I would like to thank Axel Rasche, Felix Dreher, Hendrik Hache, Christina Grimm, and

*Acknowledgments*

James Adjaye.

Without the support of several close friends it would not have been possible for me to prepare this work. First and foremost, it is my Mom who provided me invaluable support in all circumstances in life. Mom, thank you for that! And it is my Dad who taught me a positive attitude towards life, often required during the diverse phases of graduating. I am grateful to my brother and to all my friends who were at my side during my time as a PhD student. They are the essence of my daily life. I am thankful to my girl Pia. She is always on my side, bears with my daily moods, and now brings the next adventure to our lifes- our son.

# Appendix 1

In section 3.3.4, the ordinary least square estimate (OLS) is presented (see formula 3.27). Formula 3.27 can be derived from formula 3.26. Formula 3.26 requires to minimize the sum of squared residua $e_i$. Such a local extremum is identified by differentiating formula 3.26 with respect to $a$ and $b$, and by setting the resulting partial derivatives to 0. First, formula 3.26 is simplified as

$$\sum_{i=1}^{i_{max}} e_i^2 = \sum_{i=1}^{i_{max}} (x_{max_i} - (a + by_{max_i}))^2$$

$$= \sum_{i=1}^{i_{max}} (x_{max_i}^2 - 2x_{max_i}(by_{max_i} + a) + (by_{max_i} + a)^2)$$

$$= \sum_{i=1}^{i_{max}} (x_{max_i}^2 - 2by_{max_i}x_{max_i} - 2ax_{max_i} + b^2 y_{max_i}^2 + 2bay_{max_i} + a^2)$$

and the partial derivatives with respect to $a$ and $b$ are generated as follows:

$$\frac{\partial R}{\partial a} = \sum_{i=1}^{i_{max}} (-2x_{max_i} + 2by_{max_i} + 2a)$$

$$= 2 \sum_{i=1}^{i_{max}} (by_{max_i} + a - x_{max_i}) \tag{5.1}$$

$$\frac{\partial R}{\partial b} = \sum_{i=1}^{i_{max}} (-2y_{max_i}x_{max_i} + 2by_{max_i}^2 + 2ay_{max_i})$$

$$= 2 \sum_{i=1}^{i_{max}} y_{max_i}(-x_{max_i} + by_{max_i} + a) \tag{5.2}$$

By setting formula 5.1 to zero, it is valid:

$$\frac{\partial R}{\partial a} = 2 \sum_{i=1}^{i_{max}} (by_{max_i} + a - x_{max_i}) = 0$$

## Appendix 1

$$2\sum_{i=1}^{i_{max}} by_{max_i} + 2\sum_{i=1}^{i_{max}} a - 2\sum_{i=1}^{i_{max}} x_{max_i} = 0$$

$$2\sum_{i=1}^{i_{max}} by_{max_i} + 2i_{max}a - 2\sum_{i=1}^{i_{max}} x_{max_i} = 0$$

$$2i_{max}a = 2\sum_{i=1}^{i_{max}} x_{max_i} - 2\sum_{i=1}^{i_{max}} by_{max_i}$$

$$a = \frac{\sum_{i=1}^{i_{max}} x_{max_i}}{i_{max}} - \frac{\sum_{i=1}^{i_{max}} by_{max_i}}{i_{max}}$$

$$a = \frac{1}{i_{max}}\sum_{i=1}^{i_{max}} x_{max_i} - b\frac{1}{i_{max}}\sum_{i=1}^{i_{max}} y_{max_i}$$

and because of formulas 3.29 and 3.30, it is valid:

$$a = \bar{x}_{max} - b\bar{y}_{max} \tag{5.3}$$

Subsequently, formula 5.3 is introduced into formula 5.2 (i.e. the partial derivative of formula 3.26 with respect to $b$), and this partial derivative is also set to 0:

$$\frac{\partial R}{\partial b} = 2\sum_{i=1}^{i_{max}} y_{max_i}(by_{max_i} + (\bar{x}_{max} - b\bar{y}_{max}) - x_{max_i}) = 0$$

$$\sum_{i=1}^{i_{max}} y_{max_i}(by_{max_i} + (\bar{x}_{max} - b\bar{y}_{max}) - x_{max_i}) = 0$$

$$\sum_{i=1}^{i_{max}}((by_{max_i}^2 - by_{max_i}\bar{y}_{max}) - y_{max_i}x_{max_i} + y_{max_i}\bar{x}_{max}) = 0$$

$$\sum_{i=1}^{i_{max}}(by_{max_i}^2 - by_{max_i}\bar{y}_{max}) = \sum_{i=1}^{i_{max}} y_{max_i}x_{max_i} - \sum_{i=1}^{i_{max}} y_{max_i}\bar{x}_{max}$$

$$b\sum_{i=1}^{i_{max}}(y_{max_i}^2 - y_{max_i}\bar{y}_{max}) = \sum_{i=1}^{i_{max}} y_{max_i}x_{max_i} - \bar{x}_{max}\sum_{i=1}^{i_{max}} y_{max_i}$$

$$b = \frac{\sum_{i=1}^{i_{max}} y_{max_i}x_{max_i} - \bar{x}_{max}\sum_{i=1}^{i_{max}} y_{max_i}}{\sum_{i=1}^{i_{max}}(y_{max_i}^2 - y_{max_i}\bar{y}_{max})}$$

$$b = \frac{\sum_{i=1}^{i_{max}} y_{max_i}x_{max_i} - \bar{x}_{max}i_{max}\bar{y}_{max}}{\sum_{i=1}^{i_{max}} y_{max_i}^2 - \frac{1}{i_{max}}\sum_{i=1}^{i_{max}} y_{max_i}y_{max_i}} \tag{5.4}$$

The displacement law shows for the covariance

$$\sum_{i=1}^{i_{max}}(y_{max_i} - \bar{y}_{max})(x_{max_i} - \bar{x}_{max}) = \sum_{i=1}^{i_{max}}(y_{max_i}x_{max_i}) - \bar{x}_{max}i_{max}\bar{y}_{max}$$

and for the variance

## Appendix 1

$$\sum_{i=1}^{i_{max}}(y_{max_i}-\bar{y}_{max})^2 = (\sum_{i=1}^{i_{max}} y_{max_i}^2) - \frac{1}{i_{max}}(\sum_{i=1}^{i_{max}} y_{max_i})^2$$

Therefore, formula 5.4 can be notated as:

$$b = \frac{\sum_{i=1}^{i_{max}}(y_{max_i}-\bar{y}_{max})(x_{max_i}-\bar{x}_{max}))}{\sum_{i=1}^{i_{max}}(y_{max_i}-\bar{y}_{max})^2} = \frac{(S_{x_{max}y_{max}})}{(S_{y_{max}y_{max}})}$$

∎

Die VDM Verlagsservicegesellschaft sucht für wissenschaftliche Verlage abgeschlossene und herausragende

## Dissertationen, Habilitationen, Diplomarbeiten, Master Theses, Magisterarbeiten usw.

### für die kostenlose Publikation als Fachbuch.

Sie verfügen über eine Arbeit, die hohen inhaltlichen und formalen Ansprüchen genügt, und haben Interesse an einer honorarvergüteten Publikation?

Dann senden Sie bitte erste Informationen über sich und Ihre Arbeit per Email an *info@vdm-vsg.de*.

**Sie erhalten kurzfristig unser Feedback!**

VDM Verlagsservicegesellschaft mbH
Dudweiler Landstr. 99      Telefon +49 681 3720 174
D - 66123 Saarbrücken      Fax     +49 681 3720 1749
**www.vdm-vsg.de**

Die VDM Verlagsservicegesellschaft mbH vertritt

MIX
Papier aus verantwortungsvollen Quellen
Paper from responsible sources
FSC® C105338

Printed by Books on Demand GmbH, Norderstedt / Germany